Assuring Safe Operation of Robotic Systems under Uncertainty

Assuring Safe Operation of Robotic Systems under Uncertainty: Control and Learning Methods applies set-theoretic and reinforcement learning approaches to formulate, analyze, and solve the challenge of ensuring safe operation of robotic systems in an uncertain environment.

The authors adopt learning-supported, set-theoretic methods—specifically, the barrier Lyapunov function and the control barrier function—to achieve desirable robust safety with guaranteed performance in continuous-time nonlinear control applications. They also combine reinforcement learning with control theory to ensure safe learning and optimization. The reinforcement learning-based optimization framework incorporates safety and robustness guarantees by applying theoretical analysis tools from the field of control.

This book will be of interest to researchers, engineers, and students specializing in robot planning and control.

Cong Li earned a PhD from the Chair of Automatic Control Engineering, Technical University of Munich, Germany in 2022. He was also a research associate at the Chair of Automatic Control Engineering, Technical University of Munich.

Yongchao Wang is at the Xi'an Research Institution of Hi-Technology and a professor at the School of Aerospace Science and Technology, Xidian University, Xi'an, China. He was at the Chair of Automatic Control Engineering, Technical University of Munich, Germany.

Fangzhou Liu received the Doktor-Ingenieur degree in electrical engineering from the Technical University of Munich, Germany in 2019. He was a lecturer and a research fellow at the Chair of Automatic Control Engineering, Technical University of Munich, Germany. He is now a full professor at the School of Astronautics, Harbin Institute of Technology, Harbin, China.

Xinglong Zhang earned a BE in mechanical engineering from Zhejiang University, Hangzhou, China in 2011 and a PhD in system and control from the Politecnico di Milano, Italy, 2018. He is presently an associate professor at the College of Intelligence Science and Technology, National University of Defense Technology, Changsha, China. His research interests include Koopman operators, learning-based model predictive control, reinforcement learning, and approximate dynamic programming, and their applications in automotive systems.

Assuring Safe Operation of Robotic Systems under Uncertainty

Control and Learning Methods

Cong Li, Yongchao Wang, Fangzhou Liu, and Xinglong Zhang

CRC Press
Taylor & Francis Group
Boca Raton London New York

CRC Press is an imprint of the
Taylor & Francis Group, an **informa** business

This book is published with financial support from the Young Scientists Fund of the National Natural Science Foundation of China (Grant No. 62403480)

First edition published 2026
by CRC Press
2385 NW Executive Center Drive, Suite 320, Boca Raton FL 33431

and by CRC Press
4 Park Square, Milton Park, Abingdon, Oxon, OX14 4RN

CRC Press is an imprint of Taylor & Francis Group, LLC

ISBN: 978-1-041-14120-4 (hbk)
ISBN: 978-1-041-16268-1 (pbk)
ISBN: 978-1-003-68365-0 (ebk)

DOI: 10.1201/9781003683650

Typeset in Latin Modern font
by KnowledgeWorks Global Ltd.

Publisher's note: This book has been prepared from camera-ready copy provided by the authors.

To all those who have supported me throughout this journey.

Contents

SECTION II   Reinforcement Learning Approaches

CHAPTER   5 ▪ Constrained Optimal Control through Risk-Sensitive RL

CHAPTER   6 ▪ Safe Approximate Optimal Control via Filtered RL 66

CHAPTER   7 ▪ Time-Delayed Data Informed RL for Optimal Tracking Control

Preface

Autonomous systems–either robot manipulators, unmanned aerial vehicles, or self-driving cars–are desired to safely accomplish predetermined tasks with guaranteed or optimal performance despite uncertainties. Safety, often interpreted as state and input constraints that encode operational limits and actuation bounds, is critical for reliable real-world deployment. Meanwhile, uncertainties arising from unmodeled dynamics, model uncertainties, and external disturbances need to be appropriately addressed to empower autonomous systems with the adaptability toward dynamic environments. Performance objectives, such as closed-loop stability, high tracking accuracy, and minimum energy consumption, ought to be met to fulfil task requirements. This book presents the efforts applying set-theoretic and reinforcement learning approaches to formulate, analyze, and solve the aforementioned challenge, termed as safe learning control under uncertainty with guaranteed performance.

The learning-supported set-theoretic methods, specifically the barrier Lyapunov function and the control barrier function, are used in Section I (Chapter 2–Chapter 4) to achieve the desirable robust safety with guaranteed performance for continuous-time nonlinear control applications.

Safe parameter learning and control of the robot manipulator are realized in Chapter 2. This chapter online learns the uncertain dynamics of robot manipulators during the operation process, and safely improves the tracking performance gradually. The following problems are addressed simultaneously: safety issues regarding output constraints, guaranteed performance concerning tracking errors, and parametric uncertainties of robot manipulators. In particular, we adopt concurrent learning to ensure that the learned parameters converge to actual values using both real-time and historical data. Besides that, we utilize the barrier Lyapunov function to integrate safety (represented by predetermined tracking error bounds) with stability. Thereby, our designed stable control strategy based on learned dynamics could achieve safe tracking, while reducing uncertainty during the operation process at the same time.

Provably safe control under uncertainty is addressed in Chapter 3. This chapter bridges the gap between safe planning and guaranteed performance control to accomplish provable safe execution of controlled plants suffering uncertainties. This is accomplished by considering the achievable performance bound of the control level in the planning level. In particular, we first utilize time-delayed signals to formulate an uncertain and disturbed dynamics into an equivalent incremental system without using explicit model knowledge. Then, our proposed input-to-state with provable safety barrier Lyapunov function is utilized with backstepping to design a guaranteed performance tracking controller based on the above-formulated incremental system. The realizable tracking performance bound of the designed tracking controller is further considered in the planning level to generate safe reference trajectories.

Constraint learning for safe operation in unforeseen regions is discussed in Chapter 4. The preceding approaches follow a common mapping, planning, and control decoupled approach to complete safe execution under uncertainties. Alternatively, this chapter integrates perception with control levels to build safe learning systems resilient to unforeseen environments. This is achieved by the control-level quadratic optimization with the constraints, referred to as instantaneous local control barrier functions and goal-driven control Lyapunov functions, learned from perceptional signals. The integrated approach bypasses gaps among levels in the common map-plan-track decoupled paradigm to facilitate the theoretically guaranteed collision avoidance and convergence to destinations. The instantaneous local sensory data stimulates computationally cheap safe control strategies with fast adaptation to diverse uncertain environments without building a map.

The reinforcement learning and the control theory are combined in Section II (Chapter 5–Chapter 7) to achieve safe learning and optimization in the presence of uncertainties. The reinforcement learning based optimization framework is embedded with safety and robustness guarantees, applying theoretical analysis tools rooted in the control field.

Joint optimization for task performance and safety is achieved in Chapter 5. This chapter proposes an off-policy risk-sensitive reinforcement learning-based control framework to jointly optimize task performance and constraint satisfaction in a disturbed environment. The risk-sensitive state penalty terms are used to construct risk-aware value functions that penalize unsafe behaviours. The above risk-aware value function is approximated by the safety critic employing an off-policy weight update law. During the learning process, the associated approximate optimal control policy is able to satisfy both input and state constraints under disturbances.

Safe approximate optimal control is completed in Chapter 6 via filtered reinforcement learning. This chapter presents a new formulation for model-free robust constrained optimal regulation control of continuous-time nonlinear systems. The proposed RL-based approach, referred to as incremental adaptive dynamic programming, utilizes measured input-state data to allow the design of the approximate optimal incremental control strategy, stabilizing the controlled system to the target point under model uncertainties, environmental disturbances, and satisfying input saturation.

Accurate optimal tracking control is realized in Chapter 7 through time-delayed data-informed RL. To achieve safe execution in uncertain environments, the planned or replanned safe reference trajectories should be accurately tracked by a high-accuracy and robust tracking controller. Thus, this chapter investigates the optimal tracking control problem with preferences on tracking accuracy and robustness. Departing from available solutions, the developed tracking control scheme settles the curse of complexity problem in value function approximation from a decoupled way, circumvents the learning inefficiency regarding varying desired trajectories by avoiding introducing a reference trajectory dynamics into the learning process, and requires neither an accurate nor identified dynamics using time-delayed signals to facilitate model-free control.

CHAPTER 1

Introduction to Safety under Uncertainty

The provable safe execution of uncertain systems to complete predetermined tasks is required for scenarios such as robot manipulators for public services and quadrotors for search and rescue in cave environments. Both control and learning communities attempt to build paradigms to achieve provably safe control under uncertainty with guaranteed performance (considered for given missions), although with different focuses. Traditional control methods (set-theoretic methods in particular) are favored with formal guarantees of indexes such as safety and stability. However, their adaptation ability to unforeseen contexts is limited. Learning approaches (reinforcement learning specifically) allow generalization toward different environments; however, no theoretical guarantees are provided. Therefore, it is natural to bridge learning and control communities to design control schemes that enjoy rigorous theoretical guarantees and generalization toward diverse tasks and environments.

Set-Theoretic Methods.

Sets are appropriate tools to specify constraints concerning safety issues and design specifications [13]. Thus, it is attractive to investigate our interested problem in a set-theoretic context. Recently, control barrier function (CBF) has emerged as a promising set-theoretic tool to enforce safety at a control level; see [5, 36] and the references therein. Besides, the barrier Lyapunov function (BLF) [96, 95], combined with properties of barrier and Lyapunov functions, is often used with backstepping

DOI: 10.1201/9781003683650-1

to stabilize controlled plants while confining certain states into prior-given safe regions. The effectiveness of both CBF and BLF highly relies on accurate dynamics that are not always available. Toward model uncertainties, function approximation-based methods are widely utilized. The present function approximation-related works could be categorized in terms of different approximation schemes, such as polynomials [100], trigonometric series [117], orthogonal functions [8], splines [84], and neural networks (NNs) [49], etc. Among these approximation schemes, NNs play a vital role in learning-based control methods [55]. Normally, the NN approximation scheme is firstly adopted to learn a model beforehand, and then a control law is designed based on the learned model. However, the guaranteed weight convergence to the actual value is out of consideration in most of the NN approximation scheme-related works. Furthermore, the influence of unavoidable approximation errors on safety issues remains to be rigorously analyzed and addressed.

Reinforcement Learning Approaches.

Reinforcement learning (RL) provides a mathematical formulation for learning-based control strategies [93] and has shown superior performance in multiple scenarios [47, 46, 50]. Although the distinguishable model-free feature of RL overcomes the difficulty of applying traditional model-based control methods to the unknown (or hardly modeled) plants, the rigorous system stability analysis is not provided in most of the related works [79, 16]. However, a system without a stability guarantee is potentially dangerous [44]. More recently, adaptive dynamic programming (ADP) [98, 40, 99] has emerged as a promising control-theoretic RL subfield featured for available system stability proofs. ADP, implemented as an actor-critic or a critic-only NN learning structure [45, 75], forwardly solves the algebraic Riccati equation (ARE) or Hamilton-Jacobi-Bellman (HJB) equation via value function approximations. Although traditional ADP has been widely adopted to investigate stability and robustness issues, input and state constraint satisfaction during the learning process, mainly investigated for safety concerns (e.g., restrictions on torques, joint angles, and angular velocities of robot manipulators), has not yet been efficiently addressed. Violations of any constraints could lead to severe consequences such as damage to physical components. Note that the provided stability proof of ADP compromises the attractive model-free feature of RL since a mathematical form of dynamics is required to present the rigorous closed-loop stability analysis. Even though the required explicit knowledge of dynamics could be avoided by using add-on techniques such as NNs [12, 60, 109], Gaussian process (GP) [14], or observers [92], the accompanying identification processes further increase analytical complexities, computational loads, and parameter tuning efforts. Thereby, one computationally efficient and easily implemented RL-based control strategy, favored with both a model-free feature and a stability guarantee, is required.

The brief introduction illustrated above encourages us to formulate the following challenges revolving around the provable safe control under uncertainty with guaranteed performance from set-theoretic and RL perspectives.

Provable Safety under Uncertainty

Through convexing safe regions (predetermined or computed [35]) or unsafe spaces (often overly approximated), the safety problem is often investigated on the basis of accurate dynamics via tools such as CBFs [5], BLFs [96], forward (backward) reachable sets [71], model predictive control [17], penalty functions [76], and contraction theory [88], etc. The trustworthy safety checks of the safe control tools mentioned above build upon available perfect dynamics. In the case of only uncertain dynamics accessible to practitioners, current works use parametric or non-parametric methods to provide learned dynamics for safety checks. However, the gap concerning safety checks built on dynamics still in learning processes (i.e., potentially inaccurate dynamics) is not fully considered. Assumed available [88] or online/offline estimated uncertainty bounds [31] provide designers with avenues to analyze the influence of uncertainty on safety rigorously. However, the utilized conservative uncertainty bounds might reduce allowable regions into unsafe regions. This results in conservative behaviours that deteriorate performance. Therefore, rigorously quantifying, analyzing, and addressing the influence of uncertainty on safety is yet to be determined.

Guaranteed Weight Convergence

Regrading the performance issue in adaptive control, in general, the weight convergence is not required. It has been illustrated in [6, 15] that weight estimation errors may not converge to zero (indeed, they may not converge at all) even though an acceptable tracking performance is achieved. However, it is undesirable to verify safety using a constantly changing model. Thereby, the guaranteed weight convergence, offering an ensured accurate learned dynamics, is required to ensure safety under model uncertainties. Conventionally, the persistence of excitation (PE) condition is used to check the weight convergence. The weight convergence to the actual value is guaranteed if the PE condition is satisfied [15, 80]. Among existing works [28, 40], the PE condition could be satisfied by incorporating external noises to control inputs. However, this method lacks practicability given that the direct incorporation of external noises into control inputs may suffer a degradation of control performance, a waste of energy, etc. Most importantly, the incorporated external noise, ignored during the theoretical analysis process, would invalidate the provided safety and stability proofs. Hence, a fundamental problem about the guaranteed weight convergence is yet to be discussed, and one practical as well as efficient method to provide the required excitation remains to be explored.

RL for Control with Theoretical Guarantees

RL serves as one promising framework for synthesizing control policies to satisfy multiple objectives. However, RL is predominately used in simulated environments due to the lack of guaranteed safety and stability. The difficulty of providing stability proofs for RL results from noninterpretable neural network policies, unknown system dynamics, and random explorations. The control-theoretic RL method ADP relieves the stability issue via a linear approximator, a given or an identified model, and an analytical deterministic optimal control policy. However, the guaranteed safety during the learning process has not been efficiently addressed in ADP. Most of the previous ADP-related works adopt the system transformation technique to deal with state constraints [107, 29, 92]. This method, nonetheless, is limited to simple constraint forms, e.g., restricted working space in a rectangular form. Besides, although general state constraints could be tackled by the well-designed penalty functions [74, 1], which become dominant in the optimization process when possible constraint violation happens and thus punish potentially dangerous behaviours, no strict constraint satisfaction proofs are provided. However, in certain cases such as human-robot interaction scenarios, even the violation of safety-related constraints in a small possibility is unacceptable. Through the above analysis, it is meaningful to use theoretical analysis tools from the control community to inform the

RL-based policy with stability and safety guarantees. These theoretical guarantees form the basis to use RL for broad applications.

Curse of Dimensionality and Complexity

The optimal control problem is usually solved via the minimum principle of Pontryagin or dynamic programming [7]. Using dynamic programming to solve the optimal control problem faces the notorious curse of dimensionality problem, i.e., the volume of the state space grows quickly as the number of dimensions grows [77]. The RL-based ADP mitigates the curse of dimensionality problem by forwardly solving the ARE or HJB in an approximation way. However, the so-called curse of complexity appears. In particular, the number of activation functions required for the accurate value function approximation grows exponentially with the system dimension [42]. Theoretically, practitioners could seek a sufficiently large NN to achieve a satisfying approximation of a high-dimensional value function [25]. However, practically, this is nontrivial considering that appropriate activation functions are usually chosen by trial and error. This process is tedious and time-consuming. Even though a suitable set of activation functions and appropriate hyperparameters are found through engineering efforts, the accompanying computation load jeopardizes the real-time performance of the associated weight update law and control strategy [113]. Thus, experimental validations of ADP-based control strategy on a high-dimensional system are seldom found in existing works. Applying ADP to solve the optimal control problems of high-dimension systems remains to be explored.

I

Set-Theoretic Methods

CHAPTER 2

Guaranteed Safety and Performance via Concurrent Learning

2.1 PROBLEM FORMULATION

The dynamics of an n-link robot manipulator follow the Euler-Lagrange (E-L) equation:

$$M(q)\ddot{q} + C(q,\dot{q})\dot{q} + G(q) + F\dot{q} = \tau, \qquad (2.1)$$

where $M(q) : \mathbb{R}^n \to \mathbb{R}^{n\times n}$ is the symmetric positive definite inertia matrix; $C(q,\dot{q}) : \mathbb{R}^n \times \mathbb{R}^n \to \mathbb{R}^{n\times n}$ is the matrix of centrifugal and Coriolis terms; $G(q) : \mathbb{R}^n \to \mathbb{R}^n$ represents gravitational terms, and $F \in \mathbb{R}^{n\times n}$ denotes values of viscous friction; q, $\dot{q}$, and $\ddot{q} \in \mathbb{R}^n$ are the vectors of joint angles, velocities, and accelerations, respectively; $\tau \in \mathbb{R}^n$ represents the vector of input torques applied at each joint.

Property 1 *[54] The left side of the system (2.1) can be written as the following linear in parameter (LIP) form*

$$Y(q,\dot{q},\ddot{q})\theta^* = \tau, \qquad (2.2)$$

where $Y(q,\dot{q},\ddot{q}) : \mathbb{R}^n \times \mathbb{R}^n \times \mathbb{R}^n \to \mathbb{R}^{n\times m}$ is the regressor matrix; $\theta^* \in \mathbb{R}^m$ is the desired coefficient vector of the E-L equation.

Remark 2.1 *Property 1 utilizes known model properties to construct the regression matrix $Y(q,\dot{q},\ddot{q})$. The E-L equation (2.1) considers model*

DOI: 10.1201/9781003683650-2

uncertainties, such as varying joint masses, lengths, friction parameters, and unknown payloads. These uncertainties are incorporated into the coefficient vector θ^* of (2.2). For the model-free neural network (NN) approximation [49], the physical structure of the system is disregarded, often leading to sample inefficiency.

Denoting $x_1 := q$ and $x_2 := \dot{q}$, we rewrite the E-L equation (2.1) in the state-space form as

$$\begin{aligned}
\dot{x}_1 &= x_2, \\
\dot{x}_2 &= M^{-1}(x_1)(\tau - C(x_1, x_2)x_2 - G(x_1) - Fx_2), \\
y &= x_1,
\end{aligned} \qquad (2.3)$$

where $y \in \mathbb{R}^n$ is the system output that denotes the joint angles of the n-link robot manipulator (2.1), and assuming that it lies in the following set

$$\mathcal{C} = \{y \in \mathbb{R}^n : k_e \prec y \prec k_f\}. \qquad (2.4)$$

Here we consider a trajectory tracking control problem where the robot manipulator (2.1) is driven to track the desired trajectory $y_d \in \mathbb{R}^n$ precisely. Throughout this chapter, we confine ourselves that the desired trajectory y_d satisfies the following assumption.

Assumption 2.1 *The desired trajectory y_d satisfies $-\underline{y}_d \preceq y_d \preceq \overline{y}_d$, where $\underline{y}_d$ and $\overline{y}_d$ are positive constant vectors.*

The tracking error $e_1 \in \mathbb{R}^n$ is defined as

$$e_1 = y - y_d. \qquad (2.5)$$

For the safety issues considered during the trajectory tracking process, following the barrier function definition illustrated in [5], here the safety set regarding the system output y is defined as

$$\mathcal{S} = \{y \in \mathbb{R}^n : h(y) \leq 0\}, \qquad (2.6)$$

where $h(y) : \mathbb{R}^n \to \mathbb{R}$ is a continuous function. The explicit form of $h(y)$ is determined by considering various safety issues during the tracking process. As for the investigated n-link robot manipulator, considering the human-robot interactions or limited spaces, its safety set is usually defined as an allowable operation region [81, 82] that follows

$$\bar{\mathcal{S}} = \{y \in \mathbb{R}^n : k_c \prec y \prec k_d\}, \qquad (2.7)$$

where $k_c = [k_{c_1}, \ldots, k_{c_n}]^\top \in \mathbb{R}^n$ and $k_d = [k_{d_1}, \ldots, k_{d_n}]^\top \in \mathbb{R}^n$ are known constant vectors determined by controller designers. The safety set $\bar{\mathcal{S}}$ in (2.7) is a representative and explicit form of $\mathcal{S}$ in (2.6). Note that $k_c \prec -\underline{y}_d$ and $\overline{y}_d \prec k_d$ hold, i.e., y_d lies in the safety set $\bar{\mathcal{S}}$.

For the performance issues, we demand that the tracking error e_1 (2.5) lies in the performance set

$$\mathcal{P} = \{e_1 \in \mathbb{R}^n : -k_a \prec e_1 \prec k_b\}, \tag{2.8}$$

where $k_a = [k_{a_1}, \ldots, k_{a_n}]^\top \in \mathbb{R}^n$ and $k_b = [k_{b_1}, \ldots, k_{b_n}]^\top \in \mathbb{R}^n$ are predefined constant vectors.

According to (2.5), the resulting working space based on the tracking error bound (2.8) would be

$$\bar{\mathcal{P}} = \left\{y \in \mathbb{R}^n : -k_a - \underline{y}_d \prec y \prec k_b + \overline{y}_d\right\}. \tag{2.9}$$

To counter both constraints concerning safety in (2.7) and performance in (2.8), barrier Lyapunov function (BLF) [96, 95] emerges as an efficient tool. To deal with both symmetric and asymmetric constraints, a simple indicator function based BLF is proposed as

$$V(z) = p(z)\frac{z^2}{k_u^2 - z^2} + (1 - p(z))\frac{z^2}{k_l^2 - z^2}, \tag{2.10}$$

where $z \in \mathbb{R}$ is the system state; and k_l, $k_u \in \mathbb{R}$ are constraint bounds; When $z \to k_l$ or $z \to k_u$, $V(z) \to \infty$; Besides, $p(z)$ is an indicator function that follows

$$p(z) = \begin{cases} 1, & z > 0 \\ 0, & z \leq 0 \end{cases}. \tag{2.11}$$

According to Definition 2 in [96], the proposed BLF in (2.10) is an effective BLF.

For the investigated tracking control problem, the priority of safety is over performance. To improve the tracking performance while always guaranteeing safety, $\mathcal{C} \subseteq \bar{\mathcal{S}}$ and $\mathcal{C} \subseteq \bar{\mathcal{P}}$ should be satisfied together. For the purpose of achieving these considerations simultaneously, we could firstly choose the values of $-k_a$ and k_b to meet $\bar{\mathcal{P}} \subseteq \bar{\mathcal{S}}$, and then design a tracking controller to enforce $\mathcal{C} \subseteq \bar{\mathcal{P}}$. For example, consider a scenario that a robot manipulator works close to humans where the pre-planned y_d for given tasks ensures collision free with humans. To guarantee collision avoidance while accomplishing predefined tasks, we need to restrict the operation range of the robot manipulator such that $\mathcal{C} \subseteq \bar{\mathcal{S}}$, and enable the robot manipulator to track y_d precisely to satisfy $\mathcal{C} \subseteq \bar{\mathcal{P}}$, respectively. The aforementioned safety and performance requirements could be integrated together by choosing the explicit values (i.e., $-k_a$ and k_b) of the guaranteed tracking performance such that $\bar{\mathcal{P}} \subseteq \bar{\mathcal{S}}$ holds. Then, a BLF-based controller that drives the robot manipulator to track y_d with the guaranteed tracking performance also enforces the executed trajectory to lie in the restricted operation range at the same time. Although it seems to be a conservative approach, comparing to works that can only consider partial objectives of performance [10] or safety [70], the resulting BLF-based controller could drive the robot manipulator to track the desired trajectory while satisfying requirements of both safety and performance together.

Based on the aforementioned settings, the tracking control problem with guaranteed safety and performance is formulated as follows.

Problem 2.1 *Given the uncertain robot manipulator described by (2.1), and the desired trajectory y_d within the prior known safety set $\bar{\mathcal{S}}$ (2.7). Choose appropriate bounds for the performance set $\mathcal{P}$ (2.8), and design a stable adaptive control strategy based on the proposed BLF (2.10) to ensure that the manipulator tracks the desired trajectory y_d while simultaneously meeting the safety and performance requirements characterized by $\bar{\mathcal{S}}$ and $\mathcal{P}$.*

2.2 TORQUE FILTERING

For the LIP form of the E-L equation given in (2.2), measurements of the acceleration $\ddot{q}$ are required to construct the regressor matrix $Y(q, \dot{q}, \ddot{q})$. However, the information on joint acceleration is sensitive to measurement noise, which makes it unsuitable for direct use in controller design. To eliminate the need for this information, the torque filtering technique is adopted here to reformulate the original LIP form (2.2) to get a new equivalent LIP form without requirements for the joint acceleration information. Compared to the common Kalman filter that highly depends on prior knowledge (e.g., noises to be filtered) and requires extensive parameter tuning efforts [102], the adopted torque filtering technique is a simple and easily implemented method for practical applications.

To facilitate the introduction of the torque filtering technique, two auxiliary vectors $h(q, \dot{q}) : \mathbb{R}^n \times \mathbb{R}^n \to \mathbb{R}^n$ and $g(q, \dot{q}) : \mathbb{R}^n \times \mathbb{R}^n \to \mathbb{R}^n$

are firstly defined as

$$h(q, \dot{q}) = M(q)\dot{q} = Y_1(q, \dot{q})\theta^*,$$
$$g(q, \dot{q}) = -\dot{M}(q)\dot{q} + C(q, \dot{q})\dot{q} + G(q) + F\dot{q} = Y_2(q, \dot{q})\theta^*, \quad (2.12)$$

where $Y_1(q, \dot{q}) : \mathbb{R}^n \times \mathbb{R}^n \to \mathbb{R}^{n \times m}$ and $Y_2(q, \dot{q}) : \mathbb{R}^n \times \mathbb{R}^n \to \mathbb{R}^{n \times m}$ are two new regressor matrices without incorporating the information of $\ddot{q}$.

Based on the auxiliary vectors in (2.12), the system (2.1) is rewritten as

$$\dot{h}(q, \dot{q}) + g(q, \dot{q}) = (\dot{Y}_1(q, \dot{q}) + Y_2(q, \dot{q}))\theta^* = \tau, \quad (2.13)$$

where $\dot{h}(q, \dot{q}) = \dot{M}(q)\dot{q} + M(q)\ddot{q} = \dot{Y}_1(q, \dot{q})\theta^*$.

The advantage of expressing the robot manipulator in the form (2.13) is that it separates (2.1) in a way that allows $\ddot{q}$ to be filtered out. To eliminate $\ddot{q}$ from $\dot{Y}_1(q, \dot{q})$, a linear stable filter is introduced as

$$f(s) = \frac{1}{ks + 1}, \quad (2.14)$$

where s is the Laplace operator and $k \in \mathbb{R}$ is a time constant. By filtering (2.13) based on (2.14), we get the filtered version of (2.13) as

$$\dot{h}_f(q, \dot{q}) + g_f(q, \dot{q}) = \tau_f, \quad (2.15)$$

where $\dot{h}_f(q, \dot{q}) : \mathbb{R}^n \times \mathbb{R}^n \to \mathbb{R}^n$ and $g_f(q, \dot{q}) : \mathbb{R}^n \times \mathbb{R}^n \to \mathbb{R}^n$ are the filtered versions of $\dot{h}(q, \dot{q})$ and $g(q, \dot{q})$, respectively. $\tau_f \in \mathbb{R}^n$ is the filtered version of τ.

Based on (2.13), the corresponding LIP form of the filtered system (2.15) reads

$$(\dot{Y}_{1_f}(q, \dot{q}) + Y_{2_f}(q, \dot{q}))\theta^* = \tau_f, \quad (2.16)$$

where $Y_{1_f}(q, \dot{q}) : \mathbb{R}^n \times \mathbb{R}^n \to \mathbb{R}^{n \times m}$ and $Y_{2_f}(q, \dot{q}) : \mathbb{R}^n \times \mathbb{R}^n \to \mathbb{R}^{n \times m}$ are the filtered versions of the regressor matrices $Y_1(q, \dot{q})$ and $Y_2(q, \dot{q})$, respectively.

For the filter (2.14), the filtered variables and their original forms satisfy the following equations

$$k\dot{Y}_{1_f}(q, \dot{q}) + Y_{1_f}(q, \dot{q}) = Y_1(q, \dot{q}), \quad Y_{1_f}(q, \dot{q})|_{t=0} = 0,$$
$$k\dot{Y}_{2_f}(q, \dot{q}) + Y_{2_f}(q, \dot{q}) = Y_2(q, \dot{q}), \quad Y_{2_f}(q, \dot{q})|_{t=0} = 0, \quad (2.17)$$
$$k\dot{\tau}_f + \tau_f = \tau, \quad \tau_f|_{t=0} = 0.$$

Substituting the first equation of (2.17) into (2.16), finally we get the filtered LIP form of the E-L equation (2.1) as

$$Y_f(q, \dot{q})\theta^* = \tau_f, \quad (2.18)$$

where $Y_f(q, \dot{q}) = (Y_1(q, \dot{q}) - Y_{1_f}(q, \dot{q}))/k + Y_{2_f}(q, \dot{q}) : \mathbb{R}^n \times \mathbb{R}^n \to \mathbb{R}^{n \times m}$ is the new filtered regressor matrix without requirements for the joint acceleration information.

Now the new filtered regressor matrix $Y_f(q, \dot{q})$ and the resulting filtered LIP form (2.18) can be adopted to identify the unknown coefficient vector θ^* without using the joint acceleration knowledge, which is detailly clarified in the next section.

2.3 CONCURRENT LEARNING

Since the ideal coefficient vector θ^* in (2.18) is unknown, and only the estimated parameter vector $\hat{\theta}$ is available, the identification problem to be addressed here is to obtain $\hat{\theta}$ online based on the system input u, the output state y, and the filtered regression matrix Y_f. The online identification of $\hat{\theta}$ is an adaptive parameter estimation problem where the PE condition needs to be satisfied before the estimated parameters converge to the desired values. Unlike common methods that introduce external noises to satisfy the PE condition, based on the LIP form in (2.18), the TF-CL method is adopted here to guarantee the parameter convergence by utilizing both real-time and historical data.

2.3.1 Parameter Estimation Update Law

Denoting the parameter estimation error as $\tilde{\theta} = \hat{\theta} - \theta^* \in \mathbb{R}^m$. Then, the corresponding model approximation error follows

$$e_f = Y_f \tilde{\theta}. \tag{2.19}$$

Define the quadratic cost of the approximation error as $V_{e_f} = \frac{1}{2} e_f^\top e_f$. Following the standard gradient descent method to minimize V_{e_f}, the adaptive parameter estimation update law is derived as follows:

$$\dot{\hat{\theta}} = -\Gamma Y_f^\top e_f, \tag{2.20}$$

where $\Gamma \in \mathbb{R}^{m \times m}$ is a constant positive definite matrix.

It is well known that the estimated $\hat{\theta}$ converges to the desired θ^*, iff the regression matrix Y_f satisfies the PE condition [94]:

$$\int_t^{t+T} Y_f^\top(\tau) Y_f(\tau) d\tau \geq \gamma I, \tag{2.21}$$

where γ, $T \in \mathbb{R}$ are appropriate positive constants. The PE condition in (2.21) could be interpreted as requirements for a degree of data richness: when the regressor matrix Y_f varies sufficiently enough over the time interval T so that the entire γ dimension parameter space is spanned, the estimated parameters are guaranteed to converge to the desired values.

Existing works usually adopt the parameter estimation update law (2.20) and introduce external noises, such as signals in sin or cos form, to satisfy the PE condition (2.21). However, the PE condition (2.21) is hard to check online whether it is satisfied or not. An online verification condition is desirable to inform practitioners that, under this condition, the estimated parameters will converge to the desired values. The PE condition essentially relates to data richness, and only current data contributes to the common parameter estimation update law (2.20). Therefore, to get rich enough data, it is natural to incorporate historical data in constructing the parameter estimation update law.

Here, we propose a parameter estimation update law using current and historical data simultaneously. The need of adding external noises to satisfy the PE condition (2.21) is avoided with the benefit of the recorded historical data. Based on the TF-CL method, the parameter estimation update law for the unknown coefficient vector θ^* is designed as

$$\dot{\hat{\theta}} = -\Gamma k_t Y_f^\top e_f - \sum_{j=1}^{P} \Gamma k_h Y_{f_j}^\top e_{f_j}, \qquad (2.22)$$

where $k_t, k_h \in \mathbb{R}^+$ are positive constant gains to trade off the relative importance between current and historical data to the parameter estimation update law. $P \in \mathbb{R}^+$ denotes the volume of the history stacks $\mathcal{H}$ and $\mathcal{E}$. The history stacks $\mathcal{H}$ and $\mathcal{E}$ are collections of historical data, where the filtered regressor matrix Y_{f_j} and the filtered approximation error e_{f_j} denote the j-th collected data of the history stacks $\mathcal{H}$ and $\mathcal{E}$, respectively.

The parameter estimation update law (2.22) contains two parts. The first part $-\Gamma k_t Y_f^\top e_f$ relates to current data, which is a common gradient descent update law to minimize the quadratic model approximation error V_{e_f}, as like (2.20). However, an update law only with the first part cannot guarantee parameter convergence. Thus, the second part $-\sum_{j=1}^{P} \Gamma k_h Y_{f_j}^\top e_{f_j}$, which is constructed by historical data, is introduced to provide the sufficient excitation required for the parameter convergence. To analyze the parameter convergence problem based on the parameter estimation update law (2.22), a rank condition is firstly clarified in Assumption 2.2.

Assumption 2.2 *Given a history stack $\mathcal{H} = [Y_{f_1}^\top, ..., Y_{f_P}^\top] \in \mathbb{R}^{m \times (n \times P)}$, where $Y_{f_j} \in \mathbb{R}^{n \times m}$ is the j-th collected data of $\mathcal{H}$, there holds $rank(\mathcal{H}\mathcal{H}^\top) = m$.*

Note that compared to the traditional PE condition in (2.21), the rank condition of a history stack $\mathcal{H}$ in Assumption 2.2 provides an index about the richness of the historical data that could be checked online. If the rank condition is satisfied, it guarantees that the estimated parameters will converge to the desired values and vice versa. Proofs for the desirable parameter convergence are given as follows.

Theorem 2.1 *Given Assumption 2.2 and the parameter estimation update law in (2.22), the parameter estimation error $\tilde{\theta}$ converges to zero asymptotically.*

Proof 2.1 *Let $V_{cl} : \mathbb{R}^m \to \mathbb{R}$ be a candidate continuously differential Lyapunov function as*

$$V_{cl} = \frac{1}{2}\tilde{\theta}^\top \Gamma^{-1}\tilde{\theta}. \tag{2.23}$$

The bound of the Lyapunov function is

$$\frac{1}{2}\lambda_{\min}(\Gamma^{-1})\left\|\tilde{\theta}\right\|^2 \le V_{cl} \le \frac{1}{2}\lambda_{\max}(\Gamma^{-1})\left\|\tilde{\theta}\right\|^2. \tag{2.24}$$

Calculating the time derivative of V_{cl} and substituting (2.22) into it yields

$$\dot{V}_{cl} = \tilde{\theta}^\top \Gamma^{-1}\dot{\tilde{\theta}} = \tilde{\theta}^\top \Gamma^{-1}\dot{\hat{\theta}} = -k_t\tilde{\theta}^\top Y_f^\top e_f - \tilde{\theta}^\top \sum_{j=1}^{P} k_h Y_{f_j}^\top e_{f_j}$$

$$= -k_t\tilde{\theta}^\top Y_f^\top Y_f\tilde{\theta} - \tilde{\theta}^\top \sum_{j=1}^{P} k_h Y_{f_j}^\top Y_{f_j}\tilde{\theta} \le -\tilde{\theta}^\top \sum_{j=1}^{P} k_h Y_{f_j}^\top Y_{f_j}\tilde{\theta} = -\tilde{\theta}^\top Q\tilde{\theta}, \tag{2.25}$$

where $Q = \sum_{j=1}^{P} k_h Y_{f_j}^\top Y_{f_j} \in \mathbb{R}^{m \times m}$. According to Assumption 2.2, Q is positive definite and $\lambda_{\min}(Q)$ is a positive constant. Thus, the following inequality holds:

$$\dot{V}_{cl} \le -\lambda_{\min}(Q)\left\|\tilde{\theta}\right\|^2. \tag{2.26}$$

It is concluded that the parameter estimation error will converge to zero asymptotically.

2.3.2 History Stack Management Algorithm

The parameter estimation update law and the corresponding convergence proof have been provided in Theorem 2.1. The premise of Theorem 2.1 is the satisfaction of the rank condition in Assumption 2.2, i.e., a history stack $\mathcal{H}$ containing sufficiently different data is needed. Besides, according to (2.26) and $Q = \mathcal{H}\mathcal{H}^{\top}$, the convergence rate of the estimated parameters is related to the minimum eigenvalues of the history stack $\mathcal{H}$, i.e., $\lambda_{\min}(\mathcal{H}\mathcal{H}^{\top})$. With the above analysis, we know that the convergence of the estimated parameters to the desired values with a fast speed equals to (a) the satisfaction of the rank condition in Assumption 2.2 and (b) the enlargement of the minimum eigenvalue $\lambda_{\min}(\mathcal{H}\mathcal{H}^{\top})$. Thus, to achieve parameter convergence with a fast speed, in our algorithm, the history stacks $\mathcal{H}$ and $\mathcal{E}$ are updated with new data points based on two criteria: one is the data threshold ε that acts as a criterion for data difference and guides the algorithm to collect enough different data to satisfy the rank condition; the other is the minimum eigenvalue of the history stack $\mathcal{H}$ that relates to the convergence rate of the estimated parameters. Note that for computational simplicity, the minimum singular value $\sigma_{\min}(\mathcal{H}\mathcal{H}^{\top})$ is replaced with $\lambda_{\min}(\mathcal{H}\mathcal{H}^{\top})$ to act as a criterion for data storage given that $\sigma_{\min}(\mathcal{H}\mathcal{H}^{\top}) = \sqrt{\lambda_{\min}(\mathcal{H}\mathcal{H}^{\top})}$.

Given the above concerns, we design Algorithm 1 to manage the history stack. Firstly, the hyperparameter data threshold ε ensures that only new data that is sufficiently different from the latest collected data will be incorporated into the history stacks $\mathcal{H}$ and $\mathcal{E}$ (lines 2 and 8). Secondly, to improve the parameter convergence speed, when $\mathcal{H}$ reaches its volume limit P, only data points that lead to an increment of the minimum singular values of the history stack $\mathcal{H}$ will be collected. As for the method proposed in [21], the same data might be used multiple times in the history stack $\mathcal{H}$ (data richness deteriorates), and the monotonic increment of the minimum singular values cannot be guaranteed (the convergence rate of the estimated parameter is discouraged). To ensure monotonic increment of the minimum singular values, in our algorithm, the newly coming data always compares with the latest data inserted into the history stack $\mathcal{H}$. Note that the history stack volume P is a hyperparameter that requires careful tuning, which requires $P \geq m$ to satisfy the rank condition in Assumption 2.2, with m denoting the dimension of the desired coefficient vector θ^{*}.

Algorithm 1 History Stack Management Algorithm

Input: Iteration number: $i \geq 1$; Data threshold: ε; Volume: P; Auxiliary
  variables: T_h, T_e; Index: $I = P$; Empty set: S; State dimension: n.
Output: History stacks $\mathcal{H}$, $\mathcal{E}$.

1: **if** $i \leq P$ **then**
2:      **if** $\|Y_f - \mathcal{H}(:, ni - n + 1 : ni)\| / \|Y_f\| \geq \varepsilon$ **then**
3:          $\mathcal{H}(:, ni - n + 1 : ni) = Y_f^\top$ in (2.18)
4:          $\mathcal{E}(:, n) = e_f$ in (2.19)
5:          $i = i + 1$
6:      **end if**
7: **else**
8:      **if** $\|Y_f - \mathcal{H}(:, nI - n + 1 : nI)\| / \|Y_f\| \geq \varepsilon$ **then**
9:          $T_h = \mathcal{H}$; $T_e = \mathcal{E}$; $V = \sigma_{\min}(\mathcal{H}\mathcal{H}^\top)$
10:         **for** $l = 1 : P$ **do**
11:            $\mathcal{H}(:, nl - n + 1 : nl) = Y_f^\top$ in (2.18)
12:            $S(l) = \sigma_{\min}(\mathcal{H}\mathcal{H}^\top)$; $\mathcal{H} = T_h$
13:         **end for**
           $[V_{\max}, I] = \max(S)$
14:         **if** $V_{\max} \geq V$ **then**
15:            $\mathcal{H}(:, nI - n + 1 : nI) = Y_f^\top$ in (2.18)
16:            $\mathcal{E}(:, I) = e_f$ in (2.19)
17:         **else**
18:            $\mathcal{H} = T_h$; $\mathcal{E} = T_e$
19:         **end if**
20:      **end if**
21: **end if**

2.4 SAFE CONTROLLER WITH GUARANTEED PERFORMANCE

This section clarifies the design of the safe controller with guaranteed performance on the basis of the identified system in the previous section and the backstepping technique and Lyapunov analysis.

We will design a control strategy that renders the time derivative of the BLF (2.10) to be always negative semi-definite. This guarantees that with a finite initial value of the BLF, the BLF value will always be bounded during the tracking process. The boundness of the BLF implies that the safety set (2.7) and the performance set (2.8) will not be transgressed, i.e., safety and performance requirements are both satisfied.

Recall the tracking error $e_1 = x_1 - y_d$ in (2.5) and define the error $e_2 = x_2 - \alpha$, where $\alpha \in \mathbb{R}^n$ is a stabilizing function to be designed.

Step 1. The following BLF candidate is chosen to design a controller:

$$V_1 = \frac{1}{2} \sum_{i=1}^{n} p(e_{1_i}) \frac{e_{1_i}^2}{k_{b_i}^2 - e_{1_i}^2} + (1 - p(e_{1_i})) \frac{e_{1_i}^2}{k_{a_i}^2 - e_{1_i}^2}. \tag{2.27}$$

Taking time derivative of V_1 yields

$$\dot{V}_1 = \sum_{i=1}^{n} p(e_{1_i}) \frac{k_{b_i}^2 e_{1_i} \dot{e}_{1_i}}{(k_{b_i}^2 - e_{1_i}^2)^2} + (1 - p(e_{1_i})) \frac{k_{a_i}^2 e_{1_i} \dot{e}_{1_i}}{(k_{a_i}^2 - e_{1_i}^2)^2}. \tag{2.28}$$

The time derivative of e_1 is

$$\dot{e}_1 = \dot{x}_1 - \dot{y}_d = x_2 - \dot{y}_d = e_2 + \alpha - \dot{y}_d. \tag{2.29}$$

Substituting (2.29) into (2.28) yields

$$\dot{V}_1 = \sum_{i=1}^{n} p(e_{1_i}) \frac{k_{b_i}^2 e_{1_i}(e_{2_i} + \alpha_i - \dot{y}_{d_i})}{(k_{b_i}^2 - e_{1_i}^2)^2} + (1 - p(e_{1_i})) \frac{k_{a_i}^2 e_{1_i}(e_{2_i} + \alpha_i - \dot{y}_{d_i})}{(k_{a_i}^2 - e_{1_i}^2)^2}, \tag{2.30}$$

where α_i and $\dot{y}_{d_i}$ are i-th dimension of α and $\dot{y}_d$, respectively.

In order to make (2.30) be negative semi-definite, the stabilizing function α is designed as

$$\alpha = \dot{y}_d - p(e_1)(k_b^\top k_b - e_1^\top e_1)^2 k_1 e_1 - (1 - p(e_1))(k_a^\top k_a - e_1^\top e_1)^2 k_1 e_1, \tag{2.31}$$

where $k_1 \in \mathbb{R}^{n \times n}$ is a diagonal matrix of positive constants, and its i-th diagonal entry is denoted as k_{1i}. Since asymmetric constraints are considered in this chapter, the last two terms of (2.31) are designed to characterize the upper and lower constraint boundaries, i.e., k_b and $-k_a$, respectively.

For simplicity, we denote $L = \sum_{i=1}^{n} p(e_{1_i}) \frac{k_{b_i}^2 e_{1_i} e_{2_i}}{(k_{b_i}^2 - e_{1_i}^2)^2} + (1 - p(e_{1_i})) \frac{k_{a_i}^2 e_{1_i} e_{2_i}}{(k_{a_i}^2 - e_{1_i}^2)^2}$. Then, combining with (2.31), we rewrite (2.30) as

$$\dot{V}_1 = -\sum_{i=1}^{n} p(e_{1_i}) k_{1_i} k_{b_i}^2 e_{1_i}^2 + (1 - p(e_{1_i})) k_{1_i} k_{a_i}^2 e_{1_i}^2 + L \tag{2.32}$$

$$= -e_1^\top k_1 [p(e) k_b^\top k_b + (1 - p(e)) k_a^\top k_a] e_1 + L = -e_1^\top K_1 e_1 + L,$$

where $K_1 = k_1[p(e)k_b^\top k_b + (1 - p(e))k_a^\top k_a] \in \mathbb{R}^{n \times n}$ is a positive definite matrix.

Step 2. We define

$$V_2 = \frac{1}{2}e_2^\top M(x_1)e_2, \qquad (2.33)$$

and choose

$$V_{blf} = V_1 + V_2. \qquad (2.34)$$

The time derivative of V_{blf} is

$$\dot{V}_{blf} = \dot{V}_1 + \dot{V}_2 = \dot{V}_1 + e_2^\top M(x_1)\dot{e}_2 + \frac{1}{2}e_2^\top \dot{M}(x_1)e_2. \qquad (2.35)$$

Combining with (2.3), the time derivative of e_2 follows

$$\dot{e}_2 = \dot{x}_2 - \dot{\alpha} = M^{-1}(x_1)(\tau - C(x_1, x_2)x_2 - G(x_1) - Fx_2) - \dot{\alpha}. \qquad (2.36)$$

Invoking (2.32), (2.35), and (2.36) yields

$$\dot{V}_{blf} = -e_1^\top K_1 e_1 + L + e_2^\top [\tau$$
$$- C(x_1, x_2)x_2 - G(x_1) - Fx_2 - M(x_1)\dot{\alpha} + \frac{1}{2}\dot{M}(x_1)e_2]. \qquad (2.37)$$

where the stabilizing function α is defined in (2.31).

If an accurate model is available, a stabilizing control law could be directly designed as

$$\tau = M(x_1)\dot{\alpha} + C(x_1, x_2)x_2 + G(x_1) + Fx_2 - k_2e_2 - (e_2^\top)^\dagger L - \frac{1}{2}\dot{M}(x_1)e_2, \qquad (2.38)$$

where $k_2 \in \mathbb{R}^{n \times n}$ is a matrix of positive constants to be designed; $(e_2^\top)^\dagger L$ is a stabilizing term, wherein $\dagger$ stands for the Moore-Penrose inverse.

Accurate model information is required in (2.38) to design a stabilizing control law, which is, however, unavailable in our problem. To provide an approximation of the unknown model information existing in the right side of (2.38), comparing the difference between (2.2) and (2.38), a double regressor matrices technique for the TF-CL is introduced here. In particular, like the regressor matrix $Y(q, \dot{q}, \ddot{q})$ that is proposed for approximation of the system (2.2), a new regressor matrix $X(x_1, x_2, \alpha, \dot{\alpha})$ is formulated to approximate the unknown model in (2.38). Based on the newly designed $X(x_1, x_2, \alpha, \dot{\alpha})$, the following approximation equation is established:

$$X(x_1, x_2, \alpha, \dot{\alpha})\theta^* = M(x_1)\dot{\alpha} + C(x_1, x_2)x_2 + G(x_1) + Fx_2 - \frac{1}{2}\dot{M}(x_1)e_2, \qquad (2.39)$$

where $X(x_1, x_2, \alpha, \dot{\alpha}) \in \mathbb{R}^{n \times m}$ is a regressor matrix constructed based on the information of x_1, x_2, α and $\dot{\alpha}$. We defer a detailed discussion of the relationship between regressor matrices $X(x_1, x_2, \alpha, \dot{\alpha})$ and $Y(q, \dot{q}, \ddot{q})$ in Remark 2.2 and focus now on the design of the parameter estimation update law for the BLF-based controller with the help of the new regressor matrix $X(x_1, x_2, \alpha, \dot{\alpha})$.

Based on (2.39), the model-based control law (2.38) is reformulated as

$$\tau = X(x_1, x_2, \alpha, \dot{\alpha})\theta^* - k_2 e_2 - (e_2^\top)^\dagger L. \tag{2.40}$$

Since θ^* is unknown, and only $\hat{\theta}$ is available, based on the double regressor matrices technique, a TF-CL aided parameter estimation update law for the BLF-based controller (2.40) is designed as

$$\dot{\hat{\theta}} = -\Gamma X^\top e_2 - \Gamma k_t Y_f^\top e_f - \sum_{j=1}^{P} \Gamma k_h Y_{f_j}^\top e_{f_j}. \tag{2.41}$$

Comparing the difference between (2.22) and (2.41), the first term of (2.41) is designed as a stabilizing term, which serves to provide the stability proof in Theorem 2.2. By adjusting the values of Γ, k_t, and k_h, the importance of each part to the parameter estimation update law is traded off.

Finally, based on the estimated parameter vector $\hat{\theta}$ from the TF-CL method, the stabilizing control law (2.40) is rewritten as

$$\tau = X(x_1, x_2, \alpha, \dot{\alpha})\hat{\theta} - k_2 e_2 - (e_2^\top)^\dagger L. \tag{2.42}$$

Remark 2.2 *Observing (2.2) and (2.39), we find that these two equations share the same coefficient vector θ^* but with different regressor matrices. The double regressor matrices technique illustrated here makes a combination of the TF-CL method and the BLF-based control strategy feasible. $Y(q, \dot{q}, \ddot{q})$ is a regressor matrix fully depends on the model structure. $X(x_1, x_2, \alpha, \dot{\alpha})$ is a regressor matrix constructed based on both model properties and the stabilizing function α.*

In the remaining part of this section, the main conclusions of this chapter and the corresponding proofs are given based on the parameter estimation update law (2.41) and the stabilizing control strategy (2.42).

Theorem 2.2 *Consider an n-link robot manipulator (2.1), the parameter estimation update law (2.41), and the control policy (2.42). Given*

Assumptions 2.1-2.2, for initial values of the system output and the tracking error lying in the safety set (2.7) and performance set (2.8), the following properties hold:

(i) *The tracking error e_1, e_2, and the parameter estimation error $\tilde{\theta}$ are stable and converge to zero asymptotically.*

(ii) *The tracking error e_1 is bounded by Ω_{e_1} defined as*

$$\Omega_{e_1} = \left\{ e_1 \in \mathbb{R}^n : -\underline{U}_{e_1} \le e_1 \le \overline{U}_{e_1} \right\} \in \mathcal{P}, \qquad (2.43)$$

where $\underline{U}_{e_1} = [\underline{U}_{e_{1_i}}, ..., \underline{U}_{e_{1_n}}]^\top \in \mathbb{R}^n$, $\underline{U}_{e_{1_i}} = k_{a_i} \sqrt{\frac{2V(0)}{1+2V(0)}}$; $\overline{U}_{e_1} = [\overline{U}_{e_{1_i}}, ..., \overline{U}_{e_{1_n}}]^\top \in \mathbb{R}^n$, $\overline{U}_{e_{1_i}} = k_{b_i} \sqrt{\frac{2V(0)}{1+2V(0)}}$; $V(0)$ is the value of the BLF at $t = 0$. Besides, e_2 remains in the compact set Ω_{e_2} defined as

$$\Omega_{e_2} = \left\{ e_2 \in \mathbb{R}^n : \|e_2\| \le \sqrt{\frac{2V(0)}{\lambda_{\min}(M)}} \right\}. \qquad (2.44)$$

(iii) *For all $t > 0$, there holds $y(t) \in \Omega_y$, where*

$$\Omega_y = \left\{ y \in \mathbb{R}^n : -\underline{U}_{e_1} - \underline{y}_d \prec y \prec \overline{U}_{e_1} + \overline{y}_d \right\} \in \bar{\mathcal{S}}. \qquad (2.45)$$

Proof 2.2 *Proof of (i): For the stability proof, let $Z = [e_1, e_2, \tilde{\theta}]^\top \in \mathbb{R}^{2n+m}$ and consider the following Lyapunov function*

$$V(Z) = V_{blf} + V_{cl}. \qquad (2.46)$$

Combining with (2.25) and (2.37), the time derivative of (2.46) yields

$$
\begin{aligned}
\dot{V}(Z) &= \dot{V}_{blf} + \dot{V}_{cl} \\
&= -e_1^\top K_1 e_1 + L + e_2^\top [\tau - C(x_1, x_2)x_2 - G(x_1) \\
&\quad - F x_2 - M(x_1)\dot{\alpha} + \frac{1}{2}\dot{M}(x_1)e_2] + \tilde{\theta}^\top \Gamma^{-1}\dot{\tilde{\theta}}
\end{aligned}
\qquad (2.47)
$$

Substituting (2.39), (2.41), *and* (2.42) *into* (2.47) *reads*

$$\dot{V}(Z) = -e_1^\top K_1 e_1 + L + e_2^\top [X\hat{\theta} - k_2 e_2 - (e_2^\top)^\dagger L - X\theta^*] + \tilde{\theta}^\top \Gamma^{-1}\dot{\hat{\theta}}$$

$$= -e_1^\top K_1 e_1 - e_2^\top k_2 e_2 + e_2^\top X\tilde{\theta} + \tilde{\theta}^\top \Gamma^{-1}[-\Gamma X^\top e_2$$

$$- \Gamma k_t Y_f^\top e_f - \sum_{j=1}^P \Gamma k_h Y_{f_j}^\top e_{f_j}]$$

$$= -e_1^\top K_1 e_1 - e_2^\top k_2 e_2 - k_t \tilde{\theta}^\top Y_f^\top Y_f \tilde{\theta} - \tilde{\theta}^\top \sum_{j=1}^P k_h Y_{f_j}^\top (Y_{f_j}\hat{\theta}_j - \tau_{f_j})$$

$$= -e_1^\top K_1 e_1 - e_2^\top k_2 e_2 - k_t \tilde{\theta}^\top Y_f^\top Y_f \tilde{\theta} - \tilde{\theta}^\top \sum_{j=1}^P k_h Y_{f_j}^\top (Y_{f_j}\hat{\theta}_j - Y_{f_j}\theta^*)$$

$$\leq -e_1^\top K_1 e_1 - e_2^\top k_2 e_2 - \tilde{\theta}^\top \sum_{j=1}^P k_h Y_{f_j}^\top Y_{f_j} \tilde{\theta}.$$

(2.48)

Let $\mathrm{diag}(K_1, K_2, Q) \in \mathbb{R}^{(2n+m)\times(2n+m)}$, *wherein* $Q = \sum_{j=1}^P k_h Y_{f_j}^\top Y_{f_j} \in \mathbb{R}^{m\times m}$, (2.48) *could be rewritten as*

$$\dot{V}(Z) \leq -Z^\top N Z \leq - \lambda_{\min}(N) \|Z\|^2, \qquad (2.49)$$

where $\lambda_{\min}(N) = \min(\lambda_{\min}(K_1), \lambda_{\min}(k_2), \lambda_{\min}(Q))$. *Finally, it is concluded that the tracking error* e_1, *the error* e_2, *and the parameter estimation error* $\tilde{\theta}$ *converge to zero asymptotically.*

Proof of (ii): *Since* $V(Z)$ *is positive definite and* $\dot{V}(Z) < 0$ *according to* (2.49), $V(Z) \leq V(Z(0))$ *establishes. From* $V(Z) = V_1(e_1) + V_2(e_2) + V_{cl}(\tilde{\theta})$ *and the fact that* $V_2(e_2)$ *and* $V_{cl}(\tilde{\theta})$ *are positive functions, it is concluded that* $V_1(e_1) < V(Z(0))$, *i.e.,* $V_1(e_1)$ *is bounded. According to the characteristics of the BLF* (2.27), *when* $e_1 \to -k_a$ *or* $e_1 \to k_b$, *we get* $V_1(e_1) \to \infty$. *Thus, the boundness of* $V_1(e_1)$ *implies that* $e_1 \neq -k_a$ *or* $e_1 \neq k_b$. *Given that* $-k_a \prec e_1(0) \prec k_b$, *it is concluded that* $-k_a \prec e_1(t) \prec k_b, \forall t > 0$. *This means that the tracking error always lies in the required performance set* (2.8). *Besides, from the analysis mentioned above, we know that* $V_1(e_1) < V(0)$. *To get the bound of* e_1, *firstly we take the i-th element of* e_1 *as an example. For* e_{1_i}, *the following inequalities establish*

$$V(0) > \begin{cases} \dfrac{e_{1_i}^2}{2(k_{b_i}^2 - e_{1_i}^2)} & 0 < e_{1_i} < k_{b_i} \\ \dfrac{e_{1_i}^2}{2(k_{a_i}^2 - e_{1_i}^2)} & -k_{a_i} < e_{1_i} < 0 \end{cases}. \qquad (2.50)$$

We represent the above (2.50) as the following equivalent form

$$
e_{1_i}^2 < \begin{cases} k_{b_i}^2 \frac{2V(0)}{1+2V(0)} & 0 < e_{1_i} < k_{b_i} \\ k_{a_i}^2 \frac{2V(0)}{1+2V(0)} & -k_{a_i} < e_{1_i} < 0 \end{cases}. \tag{2.51}
$$

From above it is concluded that for $e_{1_i} > 0$, $e_{1_i} < k_{b_i}\sqrt{\frac{2V(0)}{1+2V(0)}}$ holds, and $e_{1_i} > -k_{a_i}\sqrt{\frac{2V(0)}{1+2V(0)}}$ establishes when $e_{1_i} < 0$. Furthermore, since $\sqrt{\frac{2V(0)}{1+2V(0)}} < 1$, $-k_{a_i} < -k_{a_i}\sqrt{\frac{2V(0)}{1+2V(0)}} < e_{1_i} < k_{b_i}\sqrt{\frac{2V(0)}{1+2V(0)}} < k_{b_i}$ establishes. Consider all elements of e_1 and the performance set $\mathcal{P}$ in (2.8), (2.43) establishes.

Consider the case of e_2, since $V_2(e_2) = \frac{1}{2}e_2^\top M e_2 < V(0)$, $\|e_2\| \leq \sqrt{\frac{2V(0)}{\lambda_{\min}(M)}}$ establishes, i.e., e_2 remains in the set Ω_{e_2}.

Proof of (iii): The output follows $y = x_1 = e_1 + y_d$. According to (2.43), $-\underline{U}_{e_1} \leq e_1 \leq \overline{U}_{e_1}$ establishes. We know that $-\underline{y}_d \leq y_d \leq \overline{y}_d$ from Assumption 2.1. Thus, it is easy to get that $-\underline{U}_{e_1} - \underline{y}_d \leq y \leq \overline{U}_{e_1} + \overline{y}_d$. Since $\underline{U}_{e_1} \prec k_a$ and $\overline{U}_{e_1} \prec k_b$, $-k_a - \underline{y}_d \prec -\underline{U}_{e_1} - \underline{y}_d \prec 0$ and $0 \prec \overline{U}_{e_1} + \overline{y}_d \prec k_b + \overline{y}_d$ establishes, i.e., $\Omega_y \in \bar{\mathcal{P}}$. Since $-k_a$ and k_b are chosen to satisfy $\mathcal{P} \subseteq \bar{\mathcal{S}}$, $\Omega_y \in \bar{\mathcal{S}}$ also establishes, i.e., system outputs will not transgress the predefined safety set (2.7).

CHAPTER 3

Provable Robust Safety through Barrier Lyapunov Function

3.1 INPUT-TO-STATE STABLE WITH PROVABLE SAFETY BLF

We consider the robot described by

$$\dot{x} = f(x) + g(x)u(x) + g(x)d, \tag{3.1}$$

where $x \in \mathbb{R}^n$, $u(x) : \mathbb{R}^n \to \mathbb{R}^m$ are the system state and control input, respectively. $f(x) : \mathbb{R}^n \to \mathbb{R}^n$, $g(x) : \mathbb{R}^n \to \mathbb{R}^{n \times m}$ are bounded and locally Lipschitz. $d \in \mathbb{L}_\infty^m$ is the assumed bounded disturbance with the (essential) supremum norm $|d|_\infty := \sup |d(t)|, t \geq 0$.

As stated in [90], if the system (3.1) admits an input-to-state stable (ISS) Lyapunov function as Definition 3.1, the system (3.1) is ISS as Definition 3.2. Therefore, designers could realize the ISS control by using one ISS-Lyapunov function to perform the controller design.

Definition 3.1 (ISS-Lyapunov Function [90]) *A smooth function $V(x) : \mathbb{R}^n \to \mathbb{R}_0^+$ is an ISS-Lyapunov function for system (3.1) if there exists $\alpha_1, \alpha_2, \alpha_3, \alpha_4 \in \mathcal{K}_\infty$ such that $\forall\, x, d$*

$$\alpha_1(|x|) \leq V(x) \leq \alpha_2(|x|) \tag{3.2a}$$

$$\dot{V}(x, d) \leq -\alpha_3(|x|) + \alpha_4(|d|). \tag{3.2b}$$

Definition 3.2 (ISS [89]) *The system (3.1) is ISS if there exists $\lambda \in \mathcal{KL}$ and $\gamma \in \mathcal{K}_\infty$*

$$|x(t, x_0, d)| \leq \lambda(|x_0|, t) + \gamma(|d|_\infty), \forall\, x_0,\, d, \forall\, t \geq 0.$$

DOI: 10.1201/9781003683650-3

The Definitions 3.1-3.2 inspire us to extend the original barrier Lyapunov function (BLF) [38], which is defined on an ideal accurate dynamics $\dot{x} = f(x) + g(x)u(x)$, to the uncertainty scenario. The resulting ISS-PS-BLF formulated in Definition 3.3 is a valid ISS-Lyapunov function in Definition 3.1 given the establishment of the inequalities (3.3a), (3.3b), and (3.3c). Furthermore, (3.3d) implies that a bounded ISS-PS-BLF would confine the state x_1 into the predetermined safe region $\mathbb{S}$. Thereby, our defined ISS-PS-BLF (3.3) provides designers with an efficient tool to realize the desired input-to-state stabilization with provable safety.

Definition 3.3 (ISS-PS-BLF) *A smooth function $V(x) := V_1(x_1) + V_2(x_2) \in \mathbb{R}_0^+$, where $x := [x_1^\top, x_2^\top]^\top \in \mathbb{R}^{n_1+n_2}$, $x_1 \in \mathbb{R}^{n_1}$, $x_2 \in \mathbb{R}^{n_2}$, is an ISS-PS-BLF for the system (3.1) on the open region $\mathbb{S} := \{x_1 \in \mathbb{R}^{n_1} : -\underline{\epsilon} \prec x_1 \prec \bar{\epsilon}\}$, where $\underline{\epsilon}_i, \bar{\epsilon}_i \in \mathbb{R}^+, \forall i \in \{1, \ldots, n_1\}$, if there exist functions $\beta_i \in \mathcal{K}_\infty$, $i = 1, \ldots, 6$, such that $\forall\, x, d$*

$$\beta_1(|x_1|) \leq V_1(x_1) \leq \beta_2(|x_1|) \tag{3.3a}$$

$$\beta_3(|x_2|) \leq V_2(x_2) \leq \beta_4(|x_2|) \tag{3.3b}$$

$$\dot{V}(x, d) \leq -\beta_5(|x|) + \beta_6(|d|) \tag{3.3c}$$

$$V_1(x_1) \to \infty, \quad x_1 \to \partial\mathbb{S}. \tag{3.3d}$$

Combining with (3.3) and the result in [38], this chapter utilizes the following candidate ISS-PS-BLF:

$$V(x) := \underbrace{\frac{1}{2} \sum_{i=1}^{n_1} \left[\frac{\bar{\epsilon}_i \underline{\epsilon}_i x_{1_i}}{(\bar{\epsilon}_i - x_{1_i})(\underline{\epsilon}_i + x_{1_i})} \right]^2}_{V_1(x_1)} + \underbrace{\frac{1}{2} x_2^\top x_2}_{V_2(x_2)}, \tag{3.4}$$

to conduct the controller design.

Here, we attempt to realize the provable safe execution of uncertain robots (3.1) in obstacle-filled environments. Our solution to this nontrivial problem is our developed SP-PGC scheme: a combination with the *performance-guaranteed control* (PGC) that explicitly quantifies the control-level performance under uncertainty and the *safe planning* (SP) where the collision-free desired trajectory is planned within the consideration of the attainable performance of the utilized controllers.

We adopt the safe planning algorithms satisfying the requirement presented in Assumption 3.1.

Assumption 3.1 *The planning level outputs a collision-free desired trajectory $p_d \in \mathbb{R}^m$ lying in a safe set $\mathbb{C} := \left\{ p(t) \in \mathbb{R}^m : \underline{p}(t) \prec p(t) \prec \overline{p}(t) \right\}$, where $\underline{p}(t), \overline{p}(t) \in \mathbb{R}^m$.*

Assumption 3.1 easily holds using off-the-self planning algorithms [51] conducted based on buffered obstacles, whose buffer size is $\epsilon \in \mathbb{R}^m$, $\epsilon_i \in \mathbb{R}^+$, $\forall i \in \{1, \ldots, m\}$. The safe execution region is $\mathbb{C} := \left\{ p \in \mathbb{R}^m : \underline{p} := p_d - \epsilon \prec p \prec \overline{p} := p_d + \epsilon \right\}$. Regarding this case, the tracking error $e_1 := p - p_d \in \mathbb{R}^m$ should satisfy $e_1 \in \mathbb{E} := \left\{ e_1 \in \mathbb{R}^m : -\underline{\epsilon} := -\epsilon \prec e_1 \prec \epsilon := \overline{\epsilon} \right\}$ to achieve safety, where $\underline{\epsilon}, \overline{\epsilon} \in \mathbb{R}^m$ are lower and upper performance bounds of e_1. Alternatively, Assumption 3.1 is easily satisfied by reachable set-based algorithms [48], or corridor (funnel) based algorithms [65, 69]. In this case, $e_1 \in \mathbb{E} := \left\{ e_1 \in \mathbb{R}^m : -\underline{\epsilon} := \underline{p} - p_d \prec e_1 \prec \overline{p} - p_d := \overline{\epsilon} \right\}$ should be guaranteed to avoid collision during practical executions.

Through the aforementioned analysis, we interpret the provable safe execution under uncertainty problem as a robust performance-guaranteed tracking control problem. This problem is nontrivial given that both state and input constraints are considered under model uncertainties and environmental disturbances. We solve this nontrivial problem via our formulated incremental system in Section 3.2 and the ISS-PS-BLF facilitated controller in Section 3.3.

3.2 DATA INFORMED INCREMENTAL SYSTEM

This section utilizes time-delayed data to formulate the incremental system that equivalently describes the movement of the original robot (3.1). By doing so, no explicit model knowledge (kinematics and/or dynamics) is required. The formulated incremental system serves as the basis for the controller design process presented in Section 3.3.

In the following, we focus on the system (3.1) satisfying Assumption 3.2 to clarify the formulation of the associated time-delayed data-informed incremental system.

Assumption 3.2 *The columns $g_1, g_2, \ldots, g_m \in \mathbb{R}^n$ of the input function $g = [g_1, g_2, \ldots, g_m]$ are linearly independent.*

Remark 3.1 *Here $g(x)$ is assumed to be full column rank such that its pseudo inverse $g^\dagger$ could be expressed as a simple algebraic formula (the inverse of $g^\top(x)g(x)$ exists). This property is widely observed in many* *physical systems, such as the quadrotor presented in Example 3.1, and the robot manipulator shown in Example 3.2 fulfills such a property.*

Firstly, introducing a prior-chosen constant matrix $\bar{g} \in \mathbb{R}^{n \times m}$ and multiplying its pseudo inverse $\bar{g}^\dagger$ on (3.1), we obtain

$$\bar{g}^\dagger \dot{x} = h + u, \tag{3.5}$$

where $h := (\bar{g}^\dagger - g^\dagger)\dot{x} + g^\dagger f + d \in \mathbb{R}^n$ embodies the unknown knowledge of the robot (3.1).

Then, we use time-delayed data to estimate h as

$$\hat{h} = h_0 = \bar{g}^\dagger \dot{x}_0 - u_0, \tag{3.6}$$

where $(\bullet)_0 = (\bullet)(t - t_s)$ denotes time-delayed data, and $t_s \in \mathbb{R}^+$ is the sampling time.

Finally, substituting (3.6) into (3.5), we get the incremental system:

$$\dot{x} = \dot{x}_0 + \bar{g}\Delta u + \bar{g}\xi, \tag{3.7}$$

where $\Delta u := u - u_0 \in \mathbb{R}^n$, and $\xi := h - \hat{h} \in \mathbb{R}^n$ is the estimation error proved to be bounded and vanishing in Lemma 3.1 under the properly chosen $\bar{g}$.

Remark 3.2 *The theoretical derivation processes* (3.5)–(3.7) *mentioned above exploit time-delayed data to transform model uncertainties and external disturbances of* (3.1) *into a provably bounded estimation error ξ of* (3.7). *This is beneficial to achieve provable safety under uncertainty given that the influence of the estimation error ξ on safety could be rigorously analyzed via an ISS approach. To achieve the same goal with our work, however, related works either estimate disturbance bounds explicitly using computation-intensive methods such as Gaussian process (GP)* [31] *or directly assume a known bound of uncertainty* [112], *which often results in conservative behaviours.*

Through the processes (3.5)-(3.7), we get an equivalent form of (3.1) without using explicit model information. In the subsequent section (3.3), we use the above-formulated incremental system (3.7) and our proposed ISS-PS-BLF (3.4) together to design the robust tracking controller with guaranteed performance.

Before proceeding to the controller design process, we provide two explicit examples to clarify how to derive the associated incremental systems from the quadrotor dynamics and the robot manipulator kinematics and dynamics.

Example 3.1 (Quadrotor) *The Euler-Lagrange (E-L) equation of a quadrotor follows [68]*

$$m\ddot{\zeta} + mg_cI_z = RT_B + T_d \tag{3.8a}$$

$$J(\eta)\ddot{\eta} + C(\eta, \dot{\eta})\dot{\eta} = \tau_B + \tau_{Bd}, \tag{3.8b}$$

where $\zeta := [x, y, z]^\top \in \mathbb{R}^3$, and $\eta := [\phi, \theta, \psi]^\top \in \mathbb{R}^3$ represent the absolute linear position and Euler angles defined in the inertial frame, respectively; $m \in \mathbb{R}^+$ denotes the mass of the quadrotor; $g_c \in \mathbb{R}^+$ is the gravity constant; $I_z := [0, 0, 1]^\top$ represents a column vector; $T_B := [0, 0, T]^\top \in \mathbb{R}^3$, where $T \in \mathbb{R}$ is the thrust in the direction of the body z-axis; $\tau_B := [\tau_\phi, \tau_\theta, \tau_\psi]^\top \in \mathbb{R}^3$ denotes the torques in the direction of the corresponding body frame angles; $T_d \in \mathbb{R}^3$ and $\tau_d \in \mathbb{R}^3$ denote the external disturbance; R, $J(\eta)$, $C(\eta, \dot{\eta}) \in \mathbb{R}^{3 \times 3}$ represent the rotation matrix, Jacobian matrix, and Coriolis term, respectively. We could rewrite the above translation dynamics (3.8a) or the attitude dynamics (3.8b) as

$$\dot{x}_1 = x_2 \tag{3.9a}$$

$$\dot{x}_2 = f + gu + gd, \tag{3.9b}$$

via letting $x_1 := \zeta$ or $\eta \in \mathbb{R}^3$, $x_2 := \dot{\zeta}$ or $\dot{\eta} \in \mathbb{R}^3$, $f := -g_cI_z$ or $-J^{-1}C(\eta, \dot{\eta})\dot{\eta} \in \mathbb{R}^3$, $g := R/m$ or $J^{-1} \in \mathbb{R}^{3 \times 3}$, $u = T_B$ or $\tau_B \in \mathbb{R}^3$, $d := R^{-1}T_d$ or $\tau_{Bd} \in \mathbb{R}^3$, respectively. Applying the theoretical derivation processes (3.5)–(3.7) mentioned above on (3.9b), we get

$$\dot{x}_1 = x_2 \tag{3.10a}$$

$$\dot{x}_2 = \dot{x}_{2,0} + \bar{g}\Delta u + \bar{g}\xi, \tag{3.10b}$$

which is an equivalent representation of (3.8) but without explicit knowledge of quadrotor dynamics.

Example 3.2 (Robot Manipulator) *The Cartesian-space position $p \in \mathbb{R}^m$ of the robot manipulator end-effector is expressed as*

$$p = h(q), \tag{3.11}$$

where $q \in \mathbb{R}^n$ is the joint-space angle vector, and $h(q) : \mathbb{R}^n \to \mathbb{R}^m$ is the differential forward kinematics. Note that $m \leq n$ holds. The end-effector velocity and acceleration $\dot{p}$, $\ddot{p} \in \mathbb{R}^m$ are related to the joint velocity and acceleration $\dot{q}$, $\ddot{q} \in \mathbb{R}^n$ as

$$\dot{p} = J(q)\dot{q} \tag{3.12a}$$

$$\ddot{p} = \dot{J}(q)\dot{q} + J(q)\ddot{q}, \tag{3.12b}$$

where $J(q) := \partial h(q)/\partial q \in \mathbb{R}^{m \times n}$ is the Jacobian matrix. Besides, its dynamics follows [58]

$$M(q)\ddot{q} + C(q, \dot{q})\dot{q} + G(q) + F_v(\dot{q}) = \tau + \tau_d, \qquad (3.13)$$

where $M(q) : \mathbb{R}^n \to \mathbb{R}^{n \times n}$ is the symmetric positive definite inertia matrix; $C(q, \dot{q}) : \mathbb{R}^n \times \mathbb{R}^n \to \mathbb{R}^{n \times n}$ is the matrix of centrifugal and Coriolis terms; $G(q) : \mathbb{R}^n \to \mathbb{R}^n$ represents the gravitational term; $F_v(\dot{q}) : \mathbb{R}^n \to \mathbb{R}^n$ denotes the viscous friction; $\tau_d \in \mathbb{R}^n$ represents the external disturbance.

Substituting (3.12) into (3.13) yields

$$M_p(q)\ddot{p} + C_p(q, \dot{q})\dot{p} + G(q) + F_v(\dot{q}) = \tau + \tau_d, \qquad (3.14)$$

where $M_p(q) := M(q)J^\dagger(q) : \mathbb{R}^n \to \mathbb{R}^{n \times m}$, $C_p(q, \dot{q}) := C(q, \dot{q})J^\dagger(q) - M(q)J^\dagger(q)\dot{J}(q)J^\dagger(q) : \mathbb{R}^n \times \mathbb{R}^n \to \mathbb{R}^{n \times m}$. The pseudo inverse follows $J^\dagger(q) := (J^\top(q)J(q))^{-1}J^\top(q) : \mathbb{R}^n \to \mathbb{R}^{n \times m}$. Then, the integrated kinematics and dynamics form (3.14) could be rewritten as the form (3.10) by denoting $x_1 := p \in \mathbb{R}^m$, $x_2 := \dot{p} \in \mathbb{R}^m$, $f := -M_p^\dagger(q)(C_p(q, \dot{q})\dot{p} + G(q) + F_v(\dot{q})) \in \mathbb{R}^m$, $g := M_p^\dagger(q) \in \mathbb{R}^{m \times n}$, $u := \tau \in \mathbb{R}^n$, and $d := \tau_d \in \mathbb{R}^n$. Through the theoretical derivation processes (3.5)–(3.7), we would get one associated incremental system of the robot manipulator (in the same form as (3.10)) without using explicit information of kinematics and dynamics.

Remark 3.3 *Examples 3.1-3.2 build on the assumption that singularities are always avoided during the whole execution process for the quadrotor and the robot manipulator. The systematic method to avoid singularity is beyond the scope of this paper. Besides, we use the pseudo-inverse of the manipulator Jacobian in (3.14) to deal with the redundancy problem of the robot manipulator case.*

3.3 MODEL-FREE PERFORMANCE-GUARANTEED CONTROL

This section utilizes our proposed ISS-PS-BLF (3.4) to develop a model-free performance-guaranteed tracking controller through a recursive controller design process. The ISS-PS-BLF provides explicit quantification of realizable tracking errors. This control-level performance quantification could feedback to the planning level to refine planned trajectories accounting for actual implementation tracking errors. The recursive controller design process based on the incremental system formulated in the previous section is illustrated as follows.

Step 1: Focusing on (3.10), the position tracking error follows $e_1 := x_1 - p_d \in \mathbb{R}^m$. To ensure that the tracking error e_1 always lies in a predetermined performance bound, $e_1 \in \mathbb{E} := \{e_1(t) \in \mathbb{R}^m : -\underline{\epsilon} \prec e_1(t) \prec \bar{\epsilon}\}$ in particular, we use the following Lyapunov function

$$V_1 := \frac{1}{2} \sum_{i=1}^{m} \left[\frac{\bar{\epsilon}_i \underline{\epsilon}_i e_{1_i}}{(\bar{\epsilon}_i - e_{1_i})(\underline{\epsilon}_i + e_{1_i})} \right]^2, \tag{3.15}$$

to facilitate the controller design. The derivative of (3.15) follows

$$\dot{V}_1 = \sum_{i=1}^{m} e_{1_i} \underbrace{\frac{\bar{\epsilon}_i^3 \underline{\epsilon}_i^3 + \bar{\epsilon}_i^2 \underline{\epsilon}_i^2 e_{1_i}^2}{(\bar{\epsilon}_i - e_{1_i})^3 (\underline{\epsilon}_i + e_{1_i})^3}}_{p_i} \dot{e}_{1_i} = e_1^\top P \dot{e}_1, \tag{3.16}$$

where $P := \mathrm{diag}(p_1, p_2, \ldots, p_m) \in \mathbb{R}^{m \times m}$.

Let $e_2 := x_2 - z \in \mathbb{R}^m$, where $z \in \mathbb{R}^m$ is a stabilizing term designed later. Combining with (3.10a), the explicit form of $\dot{e}_1$ used in (3.16) follows

$$\dot{e}_1 = \dot{x}_1 - \dot{p}_d = x_2 - \dot{p}_d = e_2 + z - \dot{p}_d. \tag{3.17}$$

Then, we design $z := \dot{p}_d - P^{-1} L_1 e_1$, wherein $L_1 := \mathrm{diag}(l_{11}, l_{12}, \ldots, l_{1m}) \in \mathbb{R}^{m \times m}$, $l_{1j} \in \mathbb{R}^+$, $j = 1, \ldots, m$. Substituting (3.17) into (3.16) yields

$$\dot{V}_1 = -e_1^\top L_1 e_1 + e_1^\top P e_2. \tag{3.18}$$

Step 2: We choose the ISS-PS-BLF as $V := V_1 + V_2$, wherein the explicit of V_2 follows

$$V_2 := \frac{1}{2} e_2^\top e_2. \tag{3.19}$$

Then, combining with (3.10b) and (3.16), we get

$$\begin{aligned} \dot{V} &= \dot{V}_1 + \dot{V}_2 \\ &= -e_1^\top L_1 e_1 + e_1^\top P e_2 + e_2^\top (\dot{x}_{2,0} + \bar{g} \Delta u + \bar{g} \xi - \dot{z}). \end{aligned} \tag{3.20}$$

Finally, we develop the incremental control input as

$$\Delta u = \bar{g}^\dagger (\dot{z} - \dot{x}_{2,0} - L_2 e_2 - P e_1), \tag{3.21}$$

to input-to-state stabilize the tracking errors e_1 and e_2 to a small neighborhood around zero as proved in Theorem 3.1, wherein $L_2 := \mathrm{diag}(l_{21}, l_{22}, \ldots, l_{2m}) \in \mathbb{R}^{m \times m}$ is a positive definite matrix, $l_{2j} \in \mathbb{R}^+$,

$j = 1, \ldots, m$. Accordingly, the control input applied at the controlled plant is recovered as

$$u = u_0 + \Delta u. \tag{3.22}$$

In the following, we theoretically analyze the properties of our designed performance-guaranteed control strategy (3.22). We firstly present the rigorous proof of the bounded estimation error in Lemma 3.1. Then, the proved bounded estimation error allows us to analyze the desirable provable safety under uncertainty in Theorem 3.1.

Lemma 3.1 *Given a sufficiently high sampling rate, there exists a positive constant $\bar{\xi} \in \mathbb{R}^+$ such that $|\xi| \leq \bar{\xi}$ holds.*

Proof 3.1 *Combining with (3.5), (3.6) and (3.10), we get*

$$
\begin{aligned}
\xi = h - h_0 &= (\bar{g}^\dagger - g^\dagger)(\dot{x}_2 - \dot{x}_{2,0}) \\
&+ (g_0^\dagger - g^\dagger)\dot{x}_{2,0} + g^\dagger(f - f_0) + (g^\dagger - g_0^\dagger)f_0 + d - d_0.
\end{aligned} \tag{3.23}
$$

Besides, focusing on (3.10b), the following equation holds

$$
\begin{aligned}
\dot{x}_2 - \dot{x}_{2,0} &= f + gu + gd - f_0 - g_0 u_0 - g_0 d_0 \\
&= g\Delta u + (g - g_0)u_0 + f - f_0 + g(d - d_0) + (g - g_0)d_0.
\end{aligned} \tag{3.24}
$$

Then, substituting (3.24) into (3.23) reads

$$\xi = (\bar{g}^\dagger g - I_{n \times n})\Delta u + \delta_1, \tag{3.25}$$

where $\delta_1 := \bar{g}^\dagger(g - g_0)u_0 + \bar{g}^\dagger(f - f_0) + \bar{g}^\dagger g(d - d_0) + \bar{g}^\dagger(g - g_0)d_0 \in \mathbb{R}^n$.
For representation simplicity, let $v := \dot{z} - L_2 e_2 - P e_1$. Accordingly, $v_0 := \dot{z}_0 - L_2 e_{2,0} - P_0 e_{1,0}$. Then, invoking (3.5), (3.6), and (3.21), we get

$$
\begin{aligned}
\Delta u = \bar{g}^\dagger(v - \dot{x}_{2,0}) &= \bar{g}^\dagger v - h_0 - u_0 \\
&= \bar{g}^\dagger v - (\bar{g}^\dagger - g_0^\dagger)\dot{x}_{2,0} + g_0^\dagger f_0 - u_0 \\
&= \bar{g}^\dagger v - (\bar{g}^\dagger - g_0^\dagger)(f_0 + g_0 u_0) + g_0^\dagger f_0 - u_0 \\
&= \bar{g}^\dagger v - \bar{g}^\dagger(f_0 + g_0 u_0) \\
&= \bar{g}^\dagger(v - v_0) - \bar{g}^\dagger(\dot{x}_{2,0} - v_0).
\end{aligned} \tag{3.26}
$$

Combining (3.10b) with (3.21) yields

$$\dot{x}_2 = v + \bar{g}\xi. \tag{3.27}$$

Besides, according to (3.27), we get

$$\xi = \bar{g}^\dagger(\dot{x}_2 - v), \quad \xi_0 = \bar{g}^\dagger(\dot{x}_{2,0} - v_0). \tag{3.28}$$

Substituting (3.28) into (3.26) implies

$$\Delta u = \bar{g}^\dagger(v - v_0) - \xi_0. \tag{3.29}$$

Finally, substituting (3.29) into (3.25), we get

$$\xi = (I_{n \times n} - \bar{g}^\dagger g)\xi_0 + \delta_1 + \delta_2, \tag{3.30}$$

where $\delta_2 := (\bar{g}^\dagger g - I_{n \times n})\bar{g}^\dagger(v - v_0) \in \mathbb{R}^n$.
For theoretical analytical purpose, we rewrite (3.30) into a discrete-time domain as

$$\xi(k) = (I_{n \times n} - \bar{g}^\dagger g(k))\xi(k-1) + \delta_1(k) + \delta_2(k). \tag{3.31}$$

Given a sufficiently high sampling rate, it is reasonable to assume that there exist positive constants $\bar{\delta}_1, \bar{\delta}_2 \in \mathbb{R}^+$ such that $|\delta_1| \leq \bar{\delta}_1$ and $|\delta_2| \leq \bar{\delta}_2$ hold. We choose the value of $\bar{g}$ to satisfy $|I_{n \times n} - \bar{g}^\dagger g(k)| \leq l < 1, l \in \mathbb{R}^+$. Then, the following equation holds

$$\begin{aligned} |\xi(k)| &\leq l\,|\xi(k-1)| + \bar{\delta}_1 + l\bar{\delta}_2 \\ &\leq l^2\,|\xi(k-2)| + (l+1)(\bar{\delta}_1 + l\bar{\delta}_2) \\ &\leq \cdots \leq l^k\,|\xi(0)| + \frac{\bar{\delta}_1 + l\bar{\delta}_2}{1-l} := \bar{\xi} \end{aligned} \tag{3.32}$$

As $k \to \infty$, $\bar{\xi} \to \frac{\bar{\delta}_1 + l\bar{\delta}_2}{1-l}$.

Theorem 3.1 *Consider the system (3.10) with the controller (3.22). Given Assumption 3.1 for initial conditions lying in the safe set $\mathbb{C}$, the following properties hold:*

1) The tracking errors e_1 and e_2 are input-to-state stabilizing to a small neighborhood around zero.

2) The Cartesian position tracking error e_1 satisfies $e_1 \in \mathbb{E}$.

3) The controlled plant realizes provable safe execution $p \in \mathbb{C}$ under model uncertainties and environmental disturbances.

Proof 3.2 Proof of 1) *Substituting (3.21) into (3.20) yields*

$$
\begin{aligned}
\dot{V} &= -e_1^\top L_1 e_1 - e_2^\top L_2 e_2 + e_2^\top \bar{g}\xi \\
&= -e_1^\top L_1 e_1 - e_2^\top (L_2 - I_{m\times m})e_2 - (e_2^\top e_2 - e_2^\top \bar{g}\xi) \\
&= -e_1^\top L_1 e_1 - e_2^\top (L_2 - I_{m\times m})e_2 - \left| e_2 - \frac{1}{2}\bar{g}\xi \right|^2 + \frac{1}{4}|\bar{g}\xi|^2 \\
&\leq -e_1^\top L_1 e_1 - e_2^\top (L_2 - I_{m\times m})e_2 + \frac{1}{4}|\bar{g}|^2 |\xi|^2 \qquad\qquad (3.33) \\
&= -e^\top L e + \frac{|\bar{g}|^2}{4}|\xi|^2 \leq -\eta_{\min}(L)|e|^2 + \frac{|\bar{g}|^2}{4}|\xi|^2 \\
&\leq -(\eta_{\min}(L) + \frac{|\bar{g}|^2}{4})|e|^2, \quad \forall |e| > |\xi|,
\end{aligned}
$$

where $e := [e_1^\top, e_2^\top]^\top \in \mathbb{R}^{2m}$, $L := \mathrm{diag}(L_1, L_2 - I_{m\times m}) \in \mathbb{R}^{2m\times 2m}$, *and the minimum eigenvalue of* L *is* $\eta_{\min}(L) := \min\{\eta_{\min}(L_1), \eta_{\min}(L_2 - I_{m\times m})\}$. *Note that* $L_2 - I_{m\times m} > 0$ *is required to make* L *as one positive definite matrix. This requirement provides practitioners with guidelines to choose suitable values of* L_2. *It is concluded that the tracking errors* e_1 *and* e_2 *are ISS with* $\alpha_3(\bullet) = -\eta_{\min}(L)|\bullet|^2$, $\alpha_4(\bullet) = \frac{|\bar{g}|^2}{4}|\bullet|^2$ *based on Definition 3.1. Then,* $|e(t)| \leq \lambda(e(t_0), t) + \gamma(|\xi(t)|_\infty)$ *holds according to Definition 3.2, i.e., the tracking error* e *remains in a ball with radius* $\lambda(e(t_0), t) + \gamma(|\xi(t)|_\infty)$. *Besides, as time* t *increases, the tracking error* e *approaches to a smaller ball of radius* $\gamma(|\xi(t)|_\infty)$ *given that for fixed* $e(t_0)$, *the* $\mathcal{KL}$ *function* λ *decreases to zero as* $t \to \infty$.

Proof of 2) *The establishment of (3.33) implies that* V *is bounded. Thereby,* V_1 *is bounded. Given that* $e_1 \to -\underline{\epsilon}$ *or* $e_1 \to \bar{\epsilon}$ *leads to* $V_1 \to \infty$ *according to (3.3d). Thus, the bounded* V_1 *proves that the tracking error* e_1 *lies in the set* $\mathbb{E}$.

Proof of 3) *The actual execution position of the controlled plant is* $p = p_d + e_1$. *Based on the fact that* $e_1 \in \mathbb{E}$, *the possible trajectory lies in the set* $\bar{\mathbb{C}} := \{p(t) \in \mathbb{R}^{n_1} : p_d - \underline{\epsilon} \prec p(t) \prec p_d + \bar{\epsilon}\}$. *By choosing* $-\underline{\epsilon} > \underline{p}(t) - p_d$ *and* $\bar{\epsilon} \prec \bar{p}(t) - p_d$ *and combining with Assumption 3.1,* $\underline{p}(t) \prec p_d - \underline{\epsilon}$ *and* $p_d + \bar{\epsilon} \prec \bar{p}(t)$ *hold. Thus, it is proved that* $\bar{\mathbb{C}} \in \mathbb{C}$, *i.e., the actual execution trajectory* $p(t)$ *always lies in the safe region* $\mathbb{C}$ *even the controlled plant (3.1) suffers from model uncertainties and environmental disturbances.*

CHAPTER 4

Safe Navigation via Integrated Perception and Control

4.1   SAFE NAVIGATION IN UNKNOWN ENVIRONMENTS

Here, we investigate the safe navigation of a mobile robot in previously unforeseen environments. The investigated mobile robot is modeled as

$$\underbrace{\begin{bmatrix} \dot{p} \\ \dot{v} \end{bmatrix}}_{\dot{x}} = \underbrace{\begin{bmatrix} 0_{2\times2} & I_{2\times2} \\ 0_{2\times2} & 0_{2\times2} \end{bmatrix} \begin{bmatrix} p \\ v \end{bmatrix}}_{f(x)} + \underbrace{\begin{bmatrix} 0_{2\times2} \\ I_{2\times2} \end{bmatrix} u}_{g(x)}, \tag{4.1}$$

where $p := [p_x, p_y]^\top$, $v := [v_x, v_y]^\top$, and $u := [u_x, u_y]^\top \in \mathbb{R}^2$ are the positions, velocities, and control inputs, respectively. For simplicity, we assume that the robot localization is perfect, i.e., the accurate vehicle state is available. The localization is realizable by the low-cost dead reckoning method. Dealing with its cumulative error is a different research direction, which is beyond the scope of this chapter.

Assume that there exist multiple prior unknown obstacles $\mathcal{O}_l$ in an environment $\mathcal{E}$, where $l \in \mathcal{L} := \{l | l = 1, 2, \ldots, L\}$ and $L \in \mathbb{N}^+$ is an uncertain value. The objective is to design a feedback controller u to drive the mobile robot (4.1) to operate safely in an uncertain environment $\mathcal{E}$ and finally reach the predetermined target position $p_d := [p_{d_x}, p_{d_y}]^\top \in \mathbb{R}^2$. We formulate the safe navigation problem mentioned above as a

DOI: 10.1201/9781003683650-4

constrained optimization problem stated as

$$\min_{u} J := \int_{t_0}^{t_f} u^\top u \, dt \tag{4.2a}$$

$$\text{s.\,t. } (4.1)$$
$$p(t_0) = p_0; v(t_0) = v_0 \tag{4.2b}$$
$$u(t) \in \mathcal{U}, \forall t \in [t_0, t_f] \tag{4.2c}$$

$$p(t) \cap \bigcup_{l=1}^{L} \mathcal{O}_l = \emptyset, \forall t \in [t_0, t_f] \tag{4.2d}$$

$$\|p(t_f) - p_d\| \leq \delta. \tag{4.2e}$$

where $\mathcal{U} \subseteq \mathbb{R}^2$ in (4.2c) denotes the bounded input space of the considered dynamics (4.1). $\delta \in \mathbb{R}^+$ in (4.2e) is a prior set threshold to check whether the reach task is completed. A quadratic control energy function is adopted in (4.2a) to reflect designers' preference on the control effort minimization.

The aforementioned safe navigation problem (4.2) is nontrivial given the constraints indicating different (might conflicting) objectives of safety and performance maximization; and the requirement of constraint satisfaction under uncertainty (limited knowledge of the environment $\mathcal{E}$). This work seeks an integrated perception and control approach to solve (4.2). In particular, we directly use perceptual inputs to learn instantaneous local control barrier functions (IL-CBFs) and goal-driven control Lyapunov functions (GD-CLFs) that are used in the control level to achieve collision avoidance and accomplish given tasks, respectively.

Before proceeding to the development of IL-CBFs and GD-CLFs, we first present the definitions of classic High-order CBF (HO-CBF) and CLF focusing on (4.1). The introduced HO-CBF and CLF here serve as theoretical basis to develop our IL-CBF and GD-CLF later.

Definition 4.1 (HO-CBF) *[105, Definition 1] Given the control system (4.1), a C^r function $h(t, x) \in \mathbb{R}$ with a relative degree r is called a (zeroing) control barrier function (of order r) if there exists a column vector $\alpha := [\alpha_1, \ldots, \alpha_r]^\top \in \mathbb{R}^r$ such that $\forall x \in \mathbb{R}^n$, $t \geq 0$,*

$$\sup_{u \in U} \left[L_g \bar{L}_f^{r-1} h(t, x) u + \bar{L}_f^r h(t, x) + \alpha^\top \xi(t, x) \right] \geq 0, \tag{4.3}$$

where $\bar{L}_f^r h := \left(\frac{\partial}{\partial t} + L_f\right)^r h$ is the modified Lie derivative of $h(t, x)$ along f and $r \in \mathbb{N}^+$, and the roots of the polynomial

$$\mathcal{P}^r(\lambda) := \lambda^r + \alpha_1 \lambda^{r-1} + \cdots + \alpha_{r-1}\lambda + \alpha_r, \tag{4.4}$$

are all negative.

Definition 4.2 (CLF) *[4, Definition 1] For the control system* (4.1), *a continuously differential function $V(x) \in \mathbb{R}$ is an exponentially stabilizing control Lyapunov function if there exists c_1, c_2, $c_3 \in \mathbb{R}^+$ such that the following equations hold*

$$c_1 \|x\|^2 \leq V(x) \leq c_2 \|x\|^2 \tag{4.5a}$$

$$\inf_{u \in \mathbb{R}^m} \left[L_f V(x) + L_g V(x)u + c_3 V(x) \right] \leq 0. \tag{4.5b}$$

4.2 IL-CBF ONLINE LEARNING

This section elucidates the mechanism of learning IL-CBFs from sensory data. In particular, the detected local obstacle information is utilized to learn the local barrier functions to describe the partial obstacle boundaries, and the learned local barrier functions update along with continuously coming data to tackle the uncertain environment. Our developed IL-CBFs are employed to formulate the QP problem in Section 4.4 to conduct collision avoidance in the control level with prior-unforeseen obstacles.

The whole boundaries of the obstacles $\mathcal{O}_l$ in $\mathcal{E}$ could be described by the barrier functions $h_l(p) \in \mathbb{R}$ using the complete knowledge of obstacles [5]. However, the obstacle information is unavailable in our investigated problem (4.2). Thus, the explicit forms of $h_l(p)$ that characterize the dangerous region $\bigcup_{l=1}^{L} \mathcal{O}_l$ are unavailable. We observe that only partial obstacle boundaries of $\mathcal{O}_l$ pose threats to the mobile robot safety at certain periods. This motivates us to utilize local sensory data to learn the local barrier functions, corresponding to the partial obstacle boundary within the mobile robot's sensor horizon, to address the collision avoidance problem.

Assume that the mobile robot is embedded with a sensor with a restricted angle S_θ and a limited horizon S_r. The value of S_θ is given, and the value of S_r satisfies

$$S_r \geq D_{\text{brake}} := \|v_{\max}\|^2 / \|a_{\max}\|, \tag{4.6}$$

where $v_{\max}$, $a_{\max} \in \mathbb{R}^2$ are the maximum velocity and breaking acceleration of the mobile robot (4.1). D_{brake} denotes the travelled distance when the mobile robot in the maximum velocity brakes using the maximum braking acceleration.

Remark 4.1 *The setting of the sensor horizon* (4.6) *is beneficial to the emergence case where our developed safe feedback control strategy fails to guarantee safety. In this scenario, the mobile robot brakes to avoid collisions.*

The sensor provides a point cloud $\mathfrak{L}$. We term $\mathfrak{D} := \{\bar{p}_1, \bar{p}_2, \ldots\} \subset \mathfrak{L}$ as the data group of the sensed obstacle boundaries, wherein $\bar{p}_i := [\bar{x}_i, \bar{y}_i]^\top \in \mathbb{R}^2$ is the position of the i-th detected obstacle boundary point. In an environment $\mathcal{E}$ with densely populated obstacles, data points in $\mathfrak{D}$ might concern multiple isolated obstacles. Therefore, we adopt the robust classifying algorithm–*density-based spatial clustering of applications with noise (DBSCAN)* [23]–to classify $\mathfrak{D}$ into multiple subgroups $\mathfrak{D}_k := \{\bar{p}_{k_1}, \bar{p}_{k_2}, \ldots\}$, wherein $\bar{p}_{k_i} := [\bar{x}_{k_i}, \bar{y}_{k_i}]^\top \in \mathbb{R}^2$ denotes the i-th data point of the k-th data group, $i \in \mathcal{I} := \{i | i = 1, \ldots, I_k\}$ with $I_k \in \mathbb{N}^+$ being the volume of $\mathfrak{D}_k$, and $k \in \mathcal{K} := \{k | k = 1, \ldots, K\}$ with $K \in \mathbb{N}^+$ being the sum of the local obstacle boundary considered in current period.

Remark 4.2 *The* DBSCAN *algorithm is compatible with our IL-CBF learning process given that it could determine the number of IL-CBFs to be learned (i.e., the values of K) automatically without using prior knowledge of environments.*

In the following, we clarify the mechanism of the IL-CBF learning focusing on the k-th data group $\mathfrak{D}_k$. Assume that i-th data pair $\bar{p}_{k_i}$ satisfies

$$\bar{y}_{k_i} = \mathcal{F}(\bar{x}_{k_i}, \zeta_k) + \varepsilon_k, \tag{4.7}$$

where $\mathcal{F}(\bar{x}_{k_i}, \zeta_k) \in \mathbb{R}$ is one n-th degree polynomial function with a parameter $\zeta_k \in \mathbb{R}^{n+1}$ to be learned; and $\varepsilon_k \sim N(0, \sigma^2)$ denotes an assumed Gaussian sensor noise with a zero mean and a constant variance $\sigma \in \mathbb{R}$.

Remark 4.3 *There exist multiple choices for $\mathcal{F}$, such as Gaussian models, linear fitting, and rational polynomials [78]. Considering the generality and simplicity issues, a polynomial model is chosen here.*

Based on (4.7) and the point cloud $\mathfrak{D}_k$ from the sensor, the parameter ζ_k is learned to minimize the approximation error:

$$\hat{\zeta}_k = \arg\min_{\zeta_k} \sum_{i=1}^{I_k} (\bar{y}_{k_i} - \mathcal{F}(\bar{x}_{k_i}, \zeta_k))^2. \tag{4.8}$$

Algorithm 2 IL-CBF Online Learning Algorithm

Input: Point cloud $\mathfrak{D}$;
Output: $\hat{h}_k, k = 1, \ldots, K$;
  1: $K = DBSCAN(\mathfrak{D})$ ▷ Robust classifying
  2: **for** $k = 1 : K$ **do**
  3:      $\hat{\zeta}_k = M\text{-}estimate(\mathfrak{D}_k)$ (4.9)   ▷ Robust regression
  4:      $\hat{h}_k = y - \mathcal{F}(x, \hat{\zeta}_k)$ (4.10)
  5: **end for**

To address potential noises and outliers that exist in the measurement data, the robust regression technique–*M-estimate* [32]–is adopted. By using the *M-estimate*, the learning of ζ_k in (4.8) is rewritten as

$$\hat{\zeta}_k = \arg\min_{\zeta_k} \sum_{i=1}^{I_k} \rho\left(\frac{\bar{y}_{k_i} - \mathcal{F}(\bar{x}_{k_i}, \zeta_k)}{\gamma}\right), \qquad (4.9)$$

where $\rho(r) = c^2/(1 - (1 - (r/c)^2)^3)$ is a robust loss function with $c = 1.345$; γ is a scale parameter estimated as $\gamma = 1.48 \left[\text{med}_i |(\bar{y}_{k_i} - \mathcal{F}(\bar{x}_{k_i}, \zeta_{k_0})) - \text{med}_i(\bar{y}_{k_i} - \mathcal{F}(\bar{x}_{k_i}, \zeta_{k_0}))|\right]$, ζ_{k_0} is the initial value of ζ_k. Details about the *M-estimate* approach are referred to [32].

Using the learned $\hat{\zeta}_k$ (4.9), we construct the IL-CBF $\hat{h}_k$ as

$$\hat{h}_k = y - \mathcal{F}(x, \hat{\zeta}_k). \qquad (4.10)$$

The IL-CBF learning process mentioned above is summarized in Algorithm 2. The mobile robot uses Algorithm 2 to update the learned IL-CBFs continuously based on the newly observed sensory data during the operation process. The IL-CBF learning is favored with computational simplicity. Thus, it is practical to update the learned IL-CBFs each step. This is favorable for the mobile robot to adapt to diverse environments.

Remark 4.4 *The clarified IL-CBF learning in this section is especially compatible with low-end sensors that only provide low-dimensional data. These limited data, however, is not enough to build a global map or describe the whole obstacle boundary.*

4.3 GD-CLF AUTOMATIC CONSTRUCTION

The data group $\mathfrak{D}$ concerning the detected obstacle boundaries is utilized in Section 4.2 to facilitate the collision avoidance in uncertain

---

Algorithm 3 GD-CLF Online Learning Algorithm

Input: Point cloud $\mathfrak{A} := \{\tilde{p}_1, \tilde{p}_2, \ldots\}$; Robot position p.
Output: $\tilde{p}_{d_j}$, and V_j, $j = 1, \ldots, J$;

1:    $\tilde{p}_{d_1} = \arg\min_{\tilde{p}_i \in \mathfrak{A}} \|\tilde{p}_i - p_d\|$ and get V_1 (4.11)
2: **if** $\|p - \tilde{p}_{d_j}\| \leq \delta$ **then**
3:      $\tilde{p}_{d_j} = \arg\min_{\tilde{p}_i \in \mathfrak{A}} \|\tilde{p}_i - p_d\|$
4:      $j = j + 1$ and update V_j (4.11)
5: **end if**

---

environments. This section exploits the remaining local collision-free sensory data group $\mathfrak{A} := \mathfrak{L} \ominus \mathfrak{D}$ to complete the long-horizon task. Specifically, we first utilize the data group $\mathfrak{A}$ to discover subgoals using a Euclidean distance metric. Then, we construct the associated GD-CLF for each subtask (subgoal). The automatically constructed GD-CLFs serve as constraints of the QP optimization in Section 4.4, whose solution ensures that the mobile robot travels toward the discovered subgoals incrementally and reaches the destination finally.

Normally, the common CLF in Definition 4.2 is inefficient to account for a long-horizon goal. Thus, through a divide-and-conquer perspective, we use sensory data $\mathfrak{A}$ to discover the subgoals $\tilde{p}_{d_j} := [x_{d_j}, y_{d_j}]^\top \in \mathbb{R}^2$, $j \in \mathcal{J} := \{j | j = 1, \ldots, J\}$ with $J \in \mathbb{N}^+$, based on a Euclidean distance metric (line 3 of Algorithm 3). In particular, we choose the nearest collision-free waypoint toward the goal position p_d as the next subgoal. These automatically determined intermediate waypoints forwardly progress toward the final desired position p_d (same with $\tilde{p}_{d_4}$).

The automatically determined subgoals $\tilde{p}_{d_j}$ from Algorithm 3 divide the long-horizon task into J short-horizon subtasks. For each subtask, we construct the GD-CLF :

$$V_j = (p - \tilde{p}_{d_j})^\top P(p - \tilde{p}_{d_j}) + (v - v_{dj})^\top Q(v - v_{dj}), j \in \mathcal{J} \qquad (4.11)$$

where P, $Q \in \mathbb{R}^{2 \times 2}$ are predetermined positive definite matrices; and $v_{d_j} \in \mathbb{R}^2$ could be a zero or a prior-given constant velocity vector. The constructed GD-CLF V_j (4.11) updates as the subgoal $\tilde{p}_{d_j}$ refreshes using Algorithm 3.

4.4 SAFE FEEDBACK CONTROL STRATEGY

This section incorporates the learned IL-CBFs (4.10) and the constructed GD-CLFs (4.11) in a QP optimization to generate the safe feedback control strategy that drives the mobile robot to safely reach the target position incrementally.

By dividing the period $[t_0, t_f]$ into multiple intervals $[t_0 + mT, t_0 + (m+1)T]$ [103], where $m \in \mathbb{N}^+$, and $T \in \mathbb{R}^+$ is the sampling time, we reformulate the original safe navigation problem (4.2) into a sequence of QPs at each interval:

$$\min_{u,\nu} \ u(t)^\top u(t) + \bar{c}_1 \nu^2(t) \tag{4.12a}$$

$$\text{s.t. } (4.1), \ (4.2b), \ (4.2c)$$

$$\ddot{\hat{h}}_k + \alpha_{k_1} \dot{\hat{h}}_k + \alpha_{k_2} \hat{h}_k \geq 0, \ k \in \mathcal{K} \tag{4.12b}$$

$$\dot{V}_j + \bar{c}_2 V_j \leq \nu, \ j \in \mathcal{J}, \tag{4.12c}$$

where $\nu(t) \in \mathbb{R}$ is a relaxation variable to relax the GD-CLF constraint to improve the QP feasibility [104]; $\alpha_{k_1}, \alpha_{k_2}, \bar{c}_1, \bar{c}_2 \in \mathbb{R}$ are parameters to be determined. The reformulated QP problem (4.12) unifies the safety requirements (4.2c), (4.12b), task requirements (4.12c), and optimization over control efforts (4.12a) to generate a multi-objective feedback controller that drives the mobile robot to progressively reach subgoals while avoiding obstacles. Note that our developed safe feedback control strategy from (4.12) only requires the information of the mobile robot position p and the target position p_d to solve the safe navigation problem (4.2) in unknown environments.

4.5 OPTIMIZED ADMISSIBLE CONTROL SPACE

The potential conflicts between the constraints (4.2c), (4.12b), and (4.12c) might result in the infeasibility problem of the QP (4.12) formulated in Section 4.4. This section formulates an optimization over the admissible control space (ACS) of the IL-CBF-associated constraint (4.12b) to improve the QP feasibility.

For analytical convenience, we denote the ACSs for constraints (4.12b) and (4.12c) as $\mathcal{A}_1 := \left\{ u \in \mathbb{R}^2 | \ddot{\hat{h}}_k + \alpha_{k_1} \dot{\hat{h}}_k + \alpha_{k_2} \hat{h}_k \geq 0, k \in \mathcal{K} \right\}$, and $\mathcal{A}_2 := \left\{ u \in \mathbb{R}^2 | \dot{V}_j + c_2 V_j \leq \nu \right\}$, respectively. Thereby, the shared control space concerning constraints (4.2c), (4.12b), and (4.12c) would be $\mathcal{S} = \mathcal{A}_1 \cap \mathcal{A}_2 \cap \mathcal{U}$. It is desirable that $\mathcal{S} \neq \emptyset$ always holds, i.e., the feasibility of the QP problem is always guaranteed. This is a nontrivial problem; especially multiple constraints are considered. Improving the possibility of satisfying $\mathcal{S} \neq \emptyset$ is equivalent to enlarge the volume of $\mathcal{S}$.

Given that the relationship between sets $\mathcal{A}_1$ and $\mathcal{A}_2$ is hard to be described and the volume of $\mathcal{U}$ is predetermined, we could transform the enlargement of the volume of $\mathcal{S}$ into the enlargement of the volumes of ACSs $\mathcal{A}_1$ and $\mathcal{A}_2$ independently. A relaxation variable ν has been used in (4.12c) to enlarge the volume of $\mathcal{A}_2$. In the following, we attempt to enlarge the volume of the ACS $\mathcal{A}_1$ to improve the feasibility of the QP problem (4.12). In particular, we firstly seek a criterion for the volume of the ACS $\mathcal{A}_1$ in Section 4.5.1 by investigating the relationship between sets $\mathcal{A}_1$ and $\mathcal{U}$. Then, a linear programming (LP) optimization problem is formulated in Section 4.5.2 to optimize the volume criterion found above to enlarge the volume of the ACS $\mathcal{A}_1$.

4.5.1 Criterion of ACS

The enlargement of $\mathcal{A}_1$ is equivalent to enlarge each IL-CBF $\hat{h}_k$ associated ACS that is denoted as $\mathcal{A}_{1_k} := \left\{ u \in \mathbb{R}^2 | \ddot{\hat{h}}_k + \alpha_{k_1} \dot{\hat{h}}_k + \alpha_{k_2} \hat{h}_k \geq 0 \right\}$, $k \in \mathcal{K}$. The explicit form of the learned k-th IL-CBF follows $\hat{h}_k = y - \hat{\zeta}_k^\top \Phi$, where $\Phi = [1, x, x^2, \dots, x^n]$. We substitute the explicit $\hat{h}_k$ into (4.12b) and rewrite the inequality as

$$Au_x + u_y + a_k^\top \Psi > 0, \tag{4.13}$$

where $A = \hat{\zeta}_k^\top \frac{\partial \Phi}{\partial x}$, $\alpha_k = [\alpha_{k_1}, \alpha_{k_2}]^\top$, $\Psi = \left[\hat{\zeta}_k^\top \frac{\partial \Phi}{\partial x} v_x - v_y, \hat{\zeta}_k^\top \frac{\partial^2 \Phi}{\partial x^2} v_x^2 + \hat{\zeta}_k^\top \Psi - y \right]^\top$.

Based on the reformulated (4.13), the geometric interpretations of the ACS $\mathcal{A}_{1_k}$ as well as the limited control input set $\mathcal{U}$ are determined. We found that a smaller value of $a_k^\top \Psi$ implies a larger area of the ACS $\mathcal{A}_{1_k}$. Thus, it is reasonable to choose the value of $a_k^\top \Psi$ as a metric to quantify the volume of the ACS $\mathcal{A}_{1_k}$, which is optimized in the subsequent subsection.

4.5.2 Optimization of ACS

This subsection clarifies the optimization over the metric $a_k^\top \Psi$, which is formulated as an LP:

$$\min_{\alpha_k} \quad \alpha_k^\top \Psi \tag{4.14a}$$

$$\text{s.t.} \ 0 < \alpha_{k_1}, \alpha_{k_2} < \overline{\alpha}_k \tag{4.14b}$$

$$a_{k1}^2 - 4\alpha_{k_2} \geq 0 \tag{4.14c}$$

where $\overline{\alpha}_k \in \mathbb{R}^+$ is the predetermined bound for the optimization variable. The formulated LP (4.14) is solved by the off-the-self *fmincon* solver. The core idea of the above LP is to select suitable values of α_{k_1} and α_{k_2} to minimize $\alpha_k^\top \Psi$ while respecting constraints (4.14b) and (4.14c). A decreased $\alpha_k^\top \Psi$ leads to a enlarged $\mathcal{A}_{1_k}$. Thereby, the QP feasibility is improved.

Remark 4.5 *The constraints (4.14b) and (4.14c) are the simplification of the following three constraints: (1) $a_{k1}^2 - 4\alpha_{k_2} \geq 0$; (2) $\frac{-\alpha_{k_1} + \sqrt{a_{k1}^2 - 4\alpha_{k_2}}}{2} < 0$; (3) $\frac{-\alpha_{k_1} - \sqrt{a_{k1}^2 - 4\alpha_{k_2}}}{2} < 0$. These three constraints ensure that the roots of (4.12b)'s related polynomials are all negative. These constraints ensure that the optimized parameter α_k^* leads to valid HO-CBFs in Definition 4.1.*

II

Reinforcement Learning Approaches

CHAPTER 5

Constrained Optimal Control through Risk-Sensitive RL

5.1 CONSTRAINED ROBUST STABILIZATION

Consider the continuous-time nonlinear dynamical system:

$$\dot{x} = f(x) + g(x)u(x) + k(x)d(x), \tag{5.1}$$

where $x \in \mathbb{R}^n$ and $u(x) \in \mathbb{R}^m$ are system states and inputs. $f(x) : \mathbb{R}^n \to \mathbb{R}^n$, $g(x) : \mathbb{R}^n \to \mathbb{R}^{n \times m}$ are the known drift and input dynamics, respectively. $k(x) : \mathbb{R}^n \to \mathbb{R}^{n \times r}$ represents the known differential system function. $d(x) : \mathbb{R}^n \to \mathbb{R}^r$ denotes the unknown additive disturbance. The general case that the additive disturbance is unmatched (i.e., $k(x) \neq g(x)$) is considered here. Assuming that $f(0) = 0$ and $d(0) = 0$, which means that the equilibrium point is $x = 0$.

Before proceeding, the following assumptions are provided. These assumptions are common in control-theoretic RL-related works and facilitate the theoretical analysis.

Assumption 5.1 *[2]* $f(x)+g(x)u$ *is Lipschitz continuous on a set* $\Omega \subseteq \mathbb{R}^n$ *that contains the origin, and the system is stabilizable on* Ω*. There exists* $g_M \in \mathbb{R}^+$ *such that the input dynamics is bounded by* $\|g(x)\| \leq g_M$*.*

Assumption 5.2 *[63]* *The unknown additive disturbance* $d(x)$ *is bounded by a known nonnegative function* $d_M(x)$*:* $\|d(x)\| \leq d_M(x)$*, and* $d_M(0) = 0$*.*

DOI: 10.1201/9781003683650-5

Based on the aforementioned settings, we formulate the constrained robust stabilization problem (CRSP) as follows.

Problem 5.1 (CRSP) *Given Assumptions 5.1-5.2, design a control strategy $u(x)$ to stabilize the closed-loop system (5.1) to the equilibrium point under additive disturbances $d(x)$, input saturation*

$$\mathbb{U}_j = \{u_j \in \mathbb{R} : |u_j| \leq \beta\}, j = 1, \ldots, m, \tag{5.2}$$

where $\beta \in \mathbb{R}^+$ is a known saturation bound; and state constraints

$$\mathbb{X}_i = \{x \in \mathbb{R}^n : h_i(x) < 0\}, i = 1, \ldots, n_c, \tag{5.3}$$

where $\mathbb{X}_i$ is a closed and convex set that contains the origin in its interior; $h_i(x) : \mathbb{R}^n \to \mathbb{R}$ is a known continuous function that relates with the i-th state constraint; $n_c \in \mathbb{N}^+$ denotes the number of considered state constraints.

5.1.1 Transformation to Optimal Control

Problem 5.1 consists of three sub-problems: disturbance rejection, input saturation, and state constraint. It is nontrivial for RL to directly deal with these sub-problems together [64]. Thus, in this section, with the pseudo-control technique proposed in [63, 62], reformulated risk-sensitive input penalty terms based on [2], and our newly designed risk-sensitive state penalty terms, we first transform the CRSP clarified as Problem 5.1 into an equivalent optimal control problem. Then, we attempt to solve the sub-problems mentioned above simultaneously under an optimization framework.

A. Pseudo-Control and Auxiliary System

As illustrated in [63], for a system suffering a matched disturbance, its disturbance-rejection control strategy could be designed by solving its nominal system's optimal control problem, wherein a cost function including the square of the disturbance bound is considered. For the unmatched disturbance $k(x)d(x)$ considered in (5.1), however, the above robust control design strategy cannot be directly applied. Thus, to address the unmatched $k(x)d(x)$ under an optimization framework as well, it is firstly decomposed as [62]

$$k(x)d(x) = g(x)\bar{d}(x) + h(x)d(x), \tag{5.4}$$

where $\bar{d}(x) = g^{\dagger}(x)k(x)d(x) : \mathbb{R}^n \rightarrow \mathbb{R}^m$, and $h(x) = (I - g(x)g^{\dagger}(x))k(x) : \mathbb{R}^n \rightarrow \mathbb{R}^{n \times r}$. Here $\dagger$ denotes the Moore-Penrose inverse. Then, we introduce the following auxiliary system with a pseudo-control $v(x) : \mathbb{R}^n \rightarrow \mathbb{R}^r$

$$\dot{x} = f(x) + g(x)u(x) + h(x)v(x), \tag{5.5}$$

to accomplish that both $g(x)\bar{d}(x)$ and $h(x)d(x)$ are matched disturbances with respect to the range of $g(x)$ and $h(x)$, respectively. Finally, similar to the robust control design strategy proposed in [63], by solving the optimal control problem of the auxiliary system (5.5) with a cost function including the square of the bounds of $\bar{d}(x)$ and $d(x)$, we could address the disturbance-rejection problem of the system (5.1) under an optimization framework. The corresponding rigorous proof is provided later in Theorem 5.1, and the following assumption is introduced for the later analysis.

Assumption 5.3 *[62] The continuous function $h(x)$ is bounded as $\|h(x)\| \leq h_M$; $\bar{d}(x)$ is bounded by a nonnegative function $l_M(x)$: $\left\|\bar{d}(x)\right\| \leq l_M(x)$, and $l_M(0) = 0$.*

B. Risk-Sensitive Input and State Penalty Terms

To tackle input and state constraints under an optimization framework, here we follow the idea of risk-sensitive RL, where multiple risk measures, e.g., high moment or conditional value at risk, are used to deal with constraints of Markov decision processes [85]. However, the available risk measures in the risk-sensitive RL field cannot guarantee strict constraint satisfaction and/or not efficient (or even inappropriate) to address constraints of continuous-time nonlinear systems. Thus, we propose risk-sensitive input penalty term (RS-IP) in Definition 5.1 and a risk-sensitive state penalty term (RS-SP) in Definition 5.2 as new risk measures during the learning process to enforce strict satisfaction of input and state constraints of continuous-time nonlinear systems.

Definition 5.1 (RS-IP) *A continuous and differential function $\phi(u)$ is a risk-sensitive input penalty term if it has the following properties:*
    (1) A bounded monotonic odd function with $\phi(0) = 0$;
    (2) The first-order partial derivatives of $\phi(u)$ is bounded.

Here the RS-IP term is a reformulation of the nonquadratic functional used in [2, 30] to confront input constraints.

Definition 5.2 (RS-SP) *Given the closed region $\mathbb{X}_i$, $i = 1, \ldots, n_c$, defined as (5.3), a continuous scalar function $S_i(x) : \mathbb{X}_i \to \mathbb{R}$, $i = 1, \ldots, n_c$, is a risk-sensitive state penalty term if the following properties hold:*

(1) $S_i(0) = 0$, and $S_i(x) > 0, \forall x \neq 0$;

(2) $S_i(x) \to \infty$ if x approaches $\partial \mathbb{X}_i$;

(3) For initial value $x(0) \in \text{Int} \, \mathbb{X}_i$, there exists $s \in \mathbb{R}^+$ such that $S_i(x(t)) \leq s, \forall t \geq 0$ along solutions of the dynamics.

Compared with similar works [74, 1] that use state penalty functions to tackle state constraints but without strict constraint satisfaction proofs, our proposed RS-SP term enables us to provide the strict constraint satisfaction proofs in Theorem 5.1. Here, the novel RS-SP term is inspired by the so-called barrier Lyapunov function [96]. The first point of Definition 5.2 denotes that $S_i(x)$ is an effective Lyapunov function candidate, which enables $S_i(x)$ to serve as part of the Lyapunov function for the system stability proof. The last two points imply that $\inf_{x \to \partial \mathbb{X}_i} S_i(x) = \infty$ and $\inf_{x \in \text{Int} \, \mathbb{X}_i} S_i(x) \geq 0$, which means that $S_i(x)$ serves as a barrier certificate for an allowable operating region $\mathbb{X}_i$.

C. Optimal Control Problem

Based on the auxiliary system (5.5) and Definitions 5.1-5.2, an equivalent optimal control problem (OCP) of the CRSP in Problem 5.1 is clarified as Problem 5.2. Comparing with traditional control-theoretic RL that accomplishes partial objectives of performance, robustness, and input/state constraint satisfaction [64, 56, 2, 74], the applied problem transformation here enables us to consider such multiple objectives together.

Problem 5.2 (OCP) *Given Assumptions 5.1-5.3, consider the auxiliary system (5.5), find $u(x)$ and $v(x)$ to minimize the cost function*

$$V(x(t)) = \int_t^\infty r(x(\tau), u(x(\tau)), v(x(\tau))) \, d\tau, \tag{5.6}$$

where the utility function follows $r(x, u(x), v(x)) = r_d(x) + \rho v^\top(x) v(x) + r_c(x, u(x))$ with $\rho \in \mathbb{R}^+$, $r_d(x) = l_M^2(x) + \rho d_M^2(x)$, and $r_c(x, u(x)) = \mathcal{W}(u(x)) + \mathcal{L}(x)$. The input penalty function $\mathcal{W}(u(x))$ follows

$$\mathcal{W}(u(x)) = \sum_{j=1}^m 2 \int_0^{u_j} \beta R_j \phi^{-1}(\vartheta_j/\beta) \, d\vartheta_j, \tag{5.7}$$

where $\phi(\cdot)$ is the RS-IP term in Definition 5.1; R_j is the j-th diagonal element of a positive definite diagonal matrix $R \in \mathbb{R}^{m \times m}$. The state penalty function $\mathcal{L}(x)$ is defined as

$$\mathcal{L}(x) = x^\top Q x + \sum_{i=1}^{n_c} k_i S_i(x), \tag{5.8}$$

where $Q \in \mathbb{R}^{n \times n}$ is a positive definite matrix; k_i is the risk sensitivity parameter that follows $k_i = 1/(1 + d_i^2)$, where d_i is the distance from the state x to the boundary of $h_i(x)$; $S_i(\cdot)$ is the RS-SP term in Definition 5.2 for the i-th state constraint.

Unlike control-theoretic RL-related works [2, 74] that incorporate nonquadratic functionals to tackle input saturation but without considering control effort-related performance, $\mathcal{W}(u(x))$ in (5.7) could take into consideration of requirements for both control limits and control energy expenditures by choosing a suitable matrix R. The commonly used risk-neutral quadratic function $x^\top Q x$ [64, 56] (capturing the desired state performance) is augmented with the newly designed weighted RS-SP term $\sum_{i=1}^{n_c} k_i S_i(x)$ (addressing multiple state constraints) to construct $\mathcal{L}(x)$ in (5.8), which enables us to consider the state-related performance and constraints together. The incorporation of $S_i(x)$ into $\mathcal{L}(x)$ deteriorates the desired performance represented by $x^\top Q x$. Therefore, we propose the risk sensitivity parameter k_i, which relates with the distance from the constraint boundary, to specify the inevitable trade-off between the state-related performance and constraint satisfaction during the learning process. Note that this kind of trade-off is ignored in existing related works [74, 1].

5.1.2 Input and State Penalty Functions

The mechanism of $\mathcal{W}(u(x))$ and $\mathcal{L}(x)$ to enable the learning process to preserve performance without violating strict input/state constraint satisfaction is detailly clarified here.

A. Mechanism of Input Penalty Function $\mathcal{W}(u(x))$

By Definition 5.1, the explicit form of the RS-IP term is chosen as $\phi(\cdot) = \tanh(\cdot)$ [2, 110]. Given the inevitable trade-off between input-related performance and constraint satisfaction, $\mathcal{W}(u(x))$ is designed to address the input constraints (5.2) and approximate $u^\top \bar{R} u$ (a common desired performance criterion for control efforts) simultaneously, where

$\bar{R} \in \mathbb{R}^{m \times m}$ is a prior-chosen positive definite matrix reflecting design-ers' preferences. The mechanism of $\mathcal{W}(u(x))$ to tackle input constraints could be clarified from two perspectives. In the first perspective, input constraints are considered in a long time-horizon. $\mathcal{W}(u(x))$ in (5.7) is an integration of $\beta R_j \tanh^{-1}(u_j/\beta)$ that is denoted as z_1. When any u_j, $j = 1, \ldots, m$, approaches to the input constraint boundaries $\pm\beta$, it fol-lows that the value of $\mathcal{W}(u(x))$ will be infinity. Since the optimization process aims to minimize the cost function, the resulting optimal con-trol strategy will be away from $\pm\beta$; Otherwise, a high value of the cost function occurs. From the other perspective, according to the later result in (5.12), the resulting optimal control strategy based on $\mathcal{W}(u(x))$ is in a form of $\tanh(\cdot)$ whose boundness enforces strict satisfaction of input constraints. Consider $\mathcal{W}_j(u_j)$, the j-th summand of $\mathcal{W}(u(x))$. It follows

$$\mathcal{W}_j(u_j) = 2\beta R_j u_j \tanh^{-1}(u_j/\beta) + \beta^2 R_j \log\left(1 - u_j^2/\beta^2\right). \qquad (5.9)$$

$\mathcal{W}_j(u_j)$ approximates the desired control energy criterion $u_j^\top \bar{R} u_j$ well via adjusting the value of R_j. Based on the above discussion, we know that $\mathcal{W}(u(x))$ in (5.7) tackles input constraints while preserving performance concerning control energy expenditures.

B. Mechanism of State Penalty Function $\mathcal{L}(x)$

According to Definition 5.2, when a potential state constraint violation happens, the corresponding RS-SP term will approach to infinity. Since the optimal control strategy aims to minimize the total cost, states will be pushed away from the direction where a high value of the RS-SP based $\mathcal{L}(x)$ occurs. Thus, the state constraint violation is avoided. To satisfy Definition 5.2, we choose $S_i(x) = \log(h_i(x))$ here. Note that the explicit form of $\log(h_i(x))$ is adjusted based on given state constraints, which is exemplified later. For a better explanation of the mechanism of the RS-SP term $S_i(x)$ and the corresponding risk sensitivity param-eter k_i, we present a four-dimensional system example with the safe re-gions defined as $\mathbb{X}_1 = \{x_1, x_2 \in \mathbb{R} : h_1(x_1, x_2) = x_1^2 + x_2^2 - 1 < 0\}$ [101], $\mathbb{X}_2 = \{x_3 \in \mathbb{R} : h_2(x_3) = |x_3| - 2 < 0\}$, and $\mathbb{X}_3 = \{x_4 \in \mathbb{R} : h_3(x_4) = |x_4| - 3 < 0\}$ [107]. The corresponding RS-SP terms are designed as $S_1(x_1, x_2) = \log(\alpha(x_2)/(\alpha(x_2) - x_1^2))$ with $\alpha(x_2) = 1 - x_2^2$, $S_2(x_3) = \log(4/(4 - x_3^2))$, and $S_3(x_4) = \log(9/(9 - x_4^2))$, respectively. These RS-SP terms act as barriers at constraint boundaries and confine the states remain in the safe regions. This inherent risk-sensitive property enables us to tackle state constraints under an optimization framework. As long as initial states lie in the safe regions and the cost function is always bounded as time evolves, the subsequent state evolution will be restricted to the safe regions. We know that the role of $S_i(x)$ will be discouraged by k_i when states are far away from the boundary of $h_i(x)$. Therefore, state-related performance is maintained when no state constraint violation occurs.

5.1.3 HJB Equation of OCP

Aiming at the transformed OCP in Problem 5.2, for any admissible control policies $u, v \in \Psi(\Omega)$, where $\Psi(\Omega)$ is the admissible control set [2, Definition 1], the associated optimal cost function follows

$$V^*(x(t)) = \min_{u,v\in\Psi(\Omega)} \int_t^{\infty} r(x(\tau), u(x(\tau)), v(x(\tau)))\, d\tau, \qquad (5.10)$$

and the HJB equation satisfies

$$0 = \min_{u,v\in\Psi(\Omega)} [\nabla V^{*^T}(f(x) + g(x)u(x) + h(x)v(x)) + r(x, u(x), v(x))]. \tag{5.11}$$

Assuming that the minimum on the right side of (5.11) exits and is unique [98]. Then, the closed forms of optimal control policies $u^*(x)$ and $v^*(x)$ are obtained as [2]

$$u^*(x) = -\beta \tanh\left(\frac{1}{2\beta}R^{-1}g^{\top}(x)\nabla V^*\right), \qquad (5.12)$$

$$v^*(x) = -\frac{1}{2\rho}h^{\top}(x)\nabla V^*. \qquad (5.13)$$

5.1.4 Problem Equivalence

Here we defer a detailed explanation of the method to get the optimal control policies (5.12) and (5.13) in Section 5.2 and focus now on the proof of equivalence between Problem 5.1 and Problem 5.2. Comparing with the result provided in [63] that merely considers additive disturbances, as shown in Theorem 5.1, the additional consideration of input and state constraints further complicates the theoretical analysis.

Theorem 5.1 *Consider the system described by (5.1) and controlled by the optimal control policy (5.12). Suppose Assumptions 5.1-5.3 hold and the initial states and control inputs lie in the predefined constraint satisfying sets (5.2) and (5.3). The optimal control policy (5.12) guarantees* *robust stabilization of the system* (5.1) *without violating the input con-straint* (5.2) *and state constraint* (5.3), *if there exists a scalar* $\epsilon_s \in \mathbb{R}^+$ *such that the following inequality is satisfied*

$$\mathcal{L}(x) > 2\rho v^{*\top}(x)v^*(x) + \epsilon_s. \tag{5.14}$$

Proof 5.1 (i) Proof of stability. *As for* $V^*(x)$ *defined as* (5.10), *we know that when* $x = 0$, $V^*(x) = 0$, *and* $V^*(x) > 0$ *for* $\forall x \neq 0$. *Thus, it can serve as a Lyapunov function candidate for stability proofs.*

Taking time derivative of $V^*(x)$ *along the system* (5.1) *yields*

$$\begin{aligned}
\dot{V}^* &= \nabla V^{*\top}(f(x) + g(x)u^*(x) + k(x)d(x)) \\
&= \nabla V^{*\top}(f(x) + g(x)u^*(x) + h(x)v^*(x)) + \nabla V^{*\top}g(x)g^\dagger(x)k(x)d(x) \\
&\quad + \nabla V^{*\top}h(x)(d(x) - v^*(x)).
\end{aligned} \tag{5.15}$$

In light of (5.11), *we get*

$$\begin{aligned}
\nabla V^{*\top}(f(x) + g(x)u^*(x) + h(x)v^*(x)) &= -\mathcal{W}(u^*(x)) - \mathcal{L}(x) \\
- \rho v^{*\top}(x)v^*(x) - l_M^2(x) - \rho d_M^2(x).
\end{aligned} \tag{5.16}$$

From (5.12), *we get*

$$\nabla V^{*\top}g(x) = -2\beta R \tanh^{-1}(u^*(x)/\beta). \tag{5.17}$$

Based on (5.13), *the following equation establishes*

$$\nabla V^{*\top}h(x) = -2\rho v^*(x). \tag{5.18}$$

Substituting (5.16), (5.17), *and* (5.18) *into* (5.15) *yields*

$$\begin{aligned}
\dot{V}^* &= -\mathcal{W}(u^*(x)) - \mathcal{L}(x) - \rho v^{*\top}(x)v^*(x) - l_M^2(x) - \rho d_M^2(x) \\
&\quad - 2\beta R \tanh^{-1}(u^*(x)/\beta)g^\dagger(x)k(x)d(x) \\
&\quad - 2\rho v^{*\top}(x)d(x) + 2\rho v^{*\top}(x)v^*(x).
\end{aligned} \tag{5.19}$$

By setting $\varsigma_j = \tanh^{-1}(\tau_j/\beta)$, *we get*

$$\begin{aligned}
\mathcal{W}(u^*(x)) &= 2\beta \sum_{j=1}^m \int_0^{u_j^*} R_j \tanh^{-1}(\tau_j/\beta)\, d\tau_j \\
&= 2\beta^2 \sum_{j=1}^m \int_0^{\tanh^{-1}(u_j^*/\beta)} R_j\varsigma_j(1 - \tanh^2(\varsigma_j))\, d\varsigma_j \\
&= \beta^2 \sum_{j=1}^m R_j(\tanh^{-1}(u_j^*(x)/\beta))^2 - \epsilon_t,
\end{aligned} \tag{5.20}$$

where $\epsilon_t = 2\beta^2 \sum_{j=1}^m \int_0^{\tanh^{-1}(u_j^*(x)/\beta)} R_j \varsigma_j \tanh^2(\varsigma_j)\, d\varsigma_j$. *Based on the integral mean-value theorem, there exist a series of* $\theta_j \in [0, \tanh^{-1}(\mu_j^*(x)/\beta], j = 1, \ldots, m,$ *such that*

$$\epsilon_t = 2\beta^2 \sum_{j=1}^m R_j \tanh^{-1}(\mu_j^*(x)/\beta)\theta_j \tanh^2(\theta_j). \tag{5.21}$$

Bearing in mind the relation (5.17) and the fact $0 < \tanh^2(\theta_j) \le 1$, *it follows that*

$$\epsilon_t \le 2\beta^2 \sum_{j=1}^m R_j \tanh^{-1}(\mu_j^*(x)/\beta)\theta_j$$

$$\le 2\beta^2 \sum_{j=1}^m R_j(\tanh^{-1}(\mu_j^*(x)/\beta))^2 \tag{5.22}$$

$$= \frac{1}{2}\nabla V^{*\top} g(x) R^{-1} g^\top(x) \nabla V^*.$$

According to the definition of the admissible policy [2], V^* is finite. Moreover, there exists $w_M > 0$ such that $\|\nabla V^*\| \le \omega_M$. Based on Assumption 5.1, we could rewrite (5.22) as

$$\epsilon_t \le b_{\epsilon_t}. \tag{5.23}$$

where $b_{\epsilon_t} = \frac{1}{2}\|R^{-1}\| g_M^2 \omega_M^2$. *Based on Assumption 5.2, the following equations establish:*

$$-2\beta R \tanh^{-1}(u^*(x)/\beta)g^\dagger(x)k(x)d(x) \le \left\|\beta R \tanh^{-1}(u^*(x)/\beta)\right\|^2 + \left\|g^\dagger(x)k(x)d(x)\right\|^2$$

$$\le \beta^2 \sum_{j=1}^m R_j^2(\tanh^{-1}(u^*(x)/\beta))^2 + l_M^2(x),$$
$$\tag{5.24}$$

$$- 2\rho v^{*\top}(x)d(x) \le \rho \|v^*(x)\|^2 + \rho \|d(x)\|^2 \le \rho \|v^*(x)\|^2 + \rho d_M^2(x). \tag{5.25}$$

Substituting (5.20), (5.23), (5.24), and (5.25) into (5.19), we have

$$\dot{V}^* \le -\mathcal{L}(x) + 2\rho v^{*\top}(x)v^*(x) + b_{\epsilon_t} + \beta^2 \sum_{j=1}^m (R_j^2 - R_j)(\tanh^{-1}(u^*(x)/\beta))^2$$

$$= -\mathcal{L}(x) + 2\rho v^{*\top}(x)v^*(x) + \epsilon_s. \tag{5.26}$$

where $\epsilon_s = b_{\epsilon_t} + \beta^2 \sum_{j=1}^m (R_j^2 - R_j)(\tanh^{-1}(u^*(x)/\beta))^2$. *Thus, $\dot{V}^* < 0$ establishes, if the condition (5.14) holds. It yields that the optimal control policy $u^*(x)$ robustly stabilizes the system (5.1).*

(ii) Proof of input and state constraint satisfaction. *Denote $V^*(0)$ as the value of the Lyapunov function candidate V^* at $t = 0$. According to the definition of admissible control policies, $V^*(0)$ is a bounded function. If $\mathcal{L}(x) > 2\rho v^{*\top}(x)v^*(x) + \epsilon_s$, $\dot{V}^* < 0$ establishes, which means that $V^*(t) < V^*(0)$, $\forall t$. The boundness of $V^*(t)$ implies that state constraints will not be violated; Otherwise, $V^*(t) \to \infty$ if any state constraint violations happens according to Definition 5.2. Since the hyperbolic tangent function satisfies $-1 \leq \tanh(\cdot) \leq 1$, the optimal control policy in (5.12) follows $-\beta \leq u^*(x) \leq \beta$, i.e., inputs are confined into the safety set (5.2). The proof provided here means that the optimal control policy $u^*(x)$ for the system (5.1) guarantees satisfaction of both constraints in terms of the system states and control inputs.*

It is proven in Theorem 5.1 that the CRSP (Problem 5.1) is equivalent to the OCP (Problem 5.2) under the inequality (5.14). Thus, in order to solve the original CRSP, the current task is to obtain the optimal control law (5.12) of the transformed OCP, which is detailedly clarified in the next section.

5.2 APPROXIMATE SOLUTION TO OCP

To get the approximate solution to the OCP, instead of introducing a common actor-critic structure used in [2, 98], here we adopt a single critic structure, which enjoys lower computation complexity [30]. Furthermore, departing from traditional methods that directly add additional noises into inputs to meet the PE condition required for the NN weight convergence [98, 12], here we apply experience data to the off-policy weight update law to achieve a sufficient excitation required for the critic NN weight convergence. Additionally, an online PER algorithm and an offline experience buffer construction algorithm are proposed as principled ways to provide the sufficient rich experience data.

5.2.1 Value Function Approximation

According to the Weierstrass high-order approximation theorem [22], there exists a weighting matrix $W^* \in \mathbb{R}^N$ such that the continuous value function is approximated as

$$V^*(x) = W^{*\top}\Phi(x) + \epsilon(x), \tag{5.27}$$

for $x \in \Omega$ with Ω being a compact set, where $\Phi(x) : \mathbb{R}^n \to \mathbb{R}^N$ is the NN activation function in a polynominal form, and $\epsilon(x) \in \mathbb{R}$ is the approximation error. Denote $\nabla\Phi \in \mathbb{R}^{N\times n}$ and $\nabla\epsilon(x) \in \mathbb{R}^n$ as the partial derivatives of $\Phi(x)$ and $\epsilon(x)$, respectively. As $N \to \infty$, both $\epsilon(x)$ and $\nabla\epsilon(x)$ converge to zero uniformly. Without loss of generality, the following assumption is given.

Assumption 5.4 *[98] There exist constants $b_\epsilon, b_{\epsilon x}, b_\Phi, b_{\Phi x} \in \mathbb{R}^+$ such that $\|\epsilon(x)\| \leq b_\epsilon$, $\|\nabla\epsilon(x)\| \leq b_{\epsilon x}$, $\|\Phi(x)\| \leq b_\Phi$, and $\|\nabla\Phi(x)\| \leq b_{\Phi x}$.*

For fixed admissible control policies $u(x)$ and $v(x)$, inserting (5.27) into (5.11) yields the Lyapunov equation (LE)

$$W^{*\top}\nabla\Phi(f(x) + g(x)u(x) + h(x)v(x)) + r(x, u(x), v(x)) = \epsilon_h, \quad (5.28)$$

where the residual error follows $\epsilon_h = -(\nabla\epsilon(x))^\top(f(x) + g(x)u(x) + h(x)v(x)) \in \mathbb{R}$. According to Assumption 5.1, the system dynamics is Lipschitz. This leads to the bounded residual error, i.e., there exists $b_{\epsilon_h} \in \mathbb{R}^+$ such that $\|\epsilon_h\| \leq b_{\epsilon_h}$.

Unlike the common analysis and derivation process in control-theoretic RL-related works [64, 56], here we rewrite the NN parameterized LE (5.28) into a linear in parameter (LIP) form that reads

$$\Theta = -W^{*\top}Y + \epsilon_h, \quad (5.29)$$

where $\Theta = r(x, u(x), v(x)) \in \mathbb{R}$, and $Y = \nabla\Phi(f(x) + g(x)u(x) + h(x)v(x)) \in \mathbb{R}^N$. Note that both Θ and Y could be obtained from real-time data.

Given the LIP form and the measurable Y, Θ in (5.29), from the perspective of adaptive control, we transform the critic NN weight W^* learning into a parameter estimation problem of an LIP system, where Y and W^* are treated as the regressor matrix and the unknown parameter vector of an LIP system, respectively. This novel transformation enables us to design a simple weight update law with guaranteed weight convergence in Section 5.2.2.

5.2.2 Off-Policy Weight Update Law

The ideal critic NN weight W^* in (5.29) is approximated by an estimated weight $\hat{W}$ which satisfies the following relation

$$\hat{\Theta} = -\hat{W}^\top Y, \quad (5.30)$$

where $\hat{\Theta} \in \mathbb{R}$ is the estimated utility function. Denoting the weight estimation error as $\tilde{W} = \hat{W} - W^* \in \mathbb{R}^N$. Then, we get

$$\tilde{\Theta} = \Theta - \hat{\Theta} = \tilde{W}^\top Y + \epsilon_h. \quad (5.31)$$

To achieve $\hat{W} \to W^*$ and $\tilde{\Theta} \to \epsilon_h$, $\hat{W}$ should be updated to minimize $E = \frac{1}{2}\tilde{\Theta}^\top \tilde{\Theta}$. Furthermore, in order to guarantee the weight convergence while minimizing E, here we exploit experience data to support the online learning process. The utilized experience data could achieve the sufficient excitation required for the weight convergence. This departs from related works [98, 12] that incorporate external noises to satisfy the PE condition. Finally, we design a simple yet efficient off-policy weight update law of the critic NN that follows

$$\dot{\hat{W}} = -\Gamma k_c Y \tilde{\Theta} - \sum_{l=1}^{P} \Gamma k_e Y_l \tilde{\Theta}_l, \tag{5.32}$$

where $\tilde{\Theta} = \Theta + \hat{W}^\top Y$ according to (5.30) and (5.31). $\Gamma \in \mathbb{R}^{N \times N}$ is a constant positive definite gain matrix. $k_c, k_e \in \mathbb{R}^+$ are constant gains to balance the relative importance between current and experience data to the online learning process. $P \in \mathbb{N}^+$ is the volume of the experience buffers $\mathfrak{B}$ and $\mathfrak{E}$, i.e., the maximum number of recorded data points. $Y_l \in \mathbb{R}^N$ and $\Theta_l \in \mathbb{R}$ denote the l-th collected data of the corresponding experience buffers $\mathfrak{B}$ and $\mathfrak{E}$, respectively. Our developed critic NN weight update law (5.32) is in a different form comparing with the counterpart in well-known control-theoretic RL-related works (see [64, 56, 98] and the references therein). Our proposed weight update law (5.32) is easily implemented and enjoys guaranteed weight convergence without causing undesirable oscillations and additional control energy expenditures.

To analyze the weight convergence of the critic NN, a rank condition about the experience buffer $\mathfrak{B}$, which serves as a richness criterion of the recorded experience data, is firstly clarified in Assumption 5.5.

Assumption 5.5 *Given* $\mathfrak{B} = [Y_1, ..., Y_P] \in \mathbb{R}^{N \times P}$, *there holds* $rank(\mathfrak{B}) = N$.

Comparing with the traditional PE condition given in [94], the rank condition regarding $\mathfrak{B}$ in Assumption 5.5 provides an index about the data richness that could be checked online, which is favorable to controller designers.

Based on the aforementioned settings, the NN weight convergence proof is shown as follows.

Theorem 5.2 *Given Assumption 5.5, the weight learning error* $\tilde{W}$ *converges to a small neighborhood around zero.*

Proof 5.2 *Consider the following candidate Lyapunov function*

$$V_{er} = \frac{1}{2}\tilde{W}^\top \Gamma^{-1}\tilde{W}. \tag{5.33}$$

The time derivative of V_{er} reads

$$
\begin{aligned}
\dot{V}_{er} &= \tilde{W}^\top \Gamma^{-1}(-\Gamma k_c Y\tilde{\Theta} - \Gamma \sum_{l=1}^{P} k_e Y_l \tilde{\Theta}_l) \\
&= -k_c \tilde{W}^\top Y\tilde{\Theta} - \tilde{W}^\top \sum_{l=1}^{P} k_e Y_l \tilde{\Theta}_l \\
&\leq -\tilde{W}^\top B\tilde{W} + \tilde{W}^\top \epsilon_{er},
\end{aligned}
\tag{5.34}
$$

where $B = \sum_{l=1}^{P} k_e Y_l Y_l^\top$, and $\epsilon_{er} = -k_c Y\epsilon_h - \sum_{l=1}^{P} k_e Y_l \epsilon_{h_l}$.

The boundness of Y and ϵ_h results in bounded ϵ_{er}, i.e., there exists $b_{\epsilon_{er}} \in \mathbb{R}^+$ such that $\|\epsilon_{er}\| \leq b_{\epsilon_{er}}$. Since B is positive definite according to Assumption 5.5, (5.34) could be written as

$$\dot{V}_{er} \leq -\|\tilde{W}\|\left(\lambda_{\min}(B)\|\tilde{W}\| - b_{\epsilon_{er}}\right). \tag{5.35}$$

Therefore, $\dot{V}_{er} < 0$ if $\|\tilde{W}\| > \frac{b_{\epsilon_{er}}}{\lambda_{\min}(B)}$. Finally, it is concluded that the weight estimation error of the critic NN will converge to the residual set

$$\Omega_{\tilde{W}} = \left\{\tilde{W}\Big| \|\tilde{W}\| \leq \frac{b_{\epsilon_{er}}}{\lambda_{\min}(B)}\right\}. \tag{5.36}$$

By observing (5.36), the size of $\Omega_{\tilde{W}}$ relates with the bound of ϵ_{er}. As $N \to \infty$, we know that $\epsilon_h \to 0$ results in $\epsilon_{er} \to 0$. Then, we get $\dot{V}_{er} \leq -\lambda_{\min}(B)\|\tilde{W}\|^2$, i.e., $\tilde{W} \to 0$ exponentially as $t \to \infty$. Equivalently, it is guaranteed that $\hat{W}$ converges to W^*.

Finally, in conjugation with (5.12) and (5.13), the approximate optimal control policies are obtained as

$$\hat{u}(x) = -\beta \tanh\left(\frac{1}{2\beta}R^{-1}g^\top(x)\nabla\Phi^\top(x)\hat{W}\right), \tag{5.37}$$

$$\hat{v}(x) = -\frac{1}{2\rho}h^\top(x)\nabla\Phi^\top(x)\hat{W}. \tag{5.38}$$

In the following part, the main conclusions are provided based on the off-policy weight update law (5.32) and the approximate optimal control policies (5.37) and (5.38).

Theorem 5.3 *Consider the dynamics (5.5), the off-policy weight update law of the critic NN in (5.32), and the control policies (5.37) and (5.38). Given Assumptions 5.1-5.5, for sufficiently large N, the approximate control policies (5.37) and (5.38) stabilize the system (5.5). Moreover, the critic NN weight learning error $\tilde{W}$ is uniformly ultimately bounded.*

Proof 5.3 *Consider the following candidate Lyapunov function*

$$J = V^*(x) + \frac{1}{2}\tilde{W}^\top \Gamma^{-1}\tilde{W}. \tag{5.39}$$

Taking the time derivative of (5.39) along the system (5.5) yields

$$\dot{J} = \dot{L}_V + \dot{L}_W. \tag{5.40}$$

where $\dot{L}_V = \dot{V}^*(x)$ and $\dot{L}_W = \tilde{W}^\top \Gamma^{-1}\dot{\tilde{W}}$.
The first term $\dot{L}_V$ follows

$$\begin{aligned}
\dot{L}_V &= \nabla V^{*\top}(f(x) + g(x)\hat{u}(x) + h(x)\hat{v}(x)) \\
&= \nabla V^{*\top}(f(x) + g(x)u^*(x) + h(x)v^*(x)) \\
&\quad + \nabla V^{*\top}g(x)(\hat{u}(x) - u^*(x)) + \nabla V^{*\top}h(x)(\hat{v}(x) - v^*(x)).
\end{aligned} \tag{5.41}$$

According to (5.16), (5.17) and (5.18), (5.41) is rewritten as

$$\begin{aligned}
\dot{L}_V &= -\mathcal{L}(x) - \mathcal{W}(u^*(x)) - \rho v^*(x)^\top v^*(x) - l_M^2(x) - \rho d_M^2(x) \\
&\quad - 2\beta R \tanh^{-1}(u^*(x)/\beta)(\hat{u}(x) - u^*(x)) - 2\rho v^{*\top}(x)(\hat{v}(x) - v^*(x)).
\end{aligned} \tag{5.42}$$

Besides, we get

$$\begin{aligned}
-2\beta R \tanh^{-1}(u^*(x)/\beta)(\hat{u}(x) - u^*(x)) &\leq \beta^2 \left\| R \tanh^{-1}(u^*(x)/\beta) \right\|^2 + \|\hat{u}(x) - u^*(x)\|^2 \\
&\leq \beta^2 \sum_{j=1}^{m} R_j^2(\tanh^{-1}(u_j^*(x)/\beta))^2 + \|\hat{u}(x) - u^*(x)\|^2.
\end{aligned} \tag{5.43}$$

Based on (5.20)-(5.26), the following equation also establishes

$$\begin{aligned}
&- \mathcal{W}(u^*(x)) - 2\beta \tanh^{-1}(u^*(x)/\beta)(\hat{u}(x) - u^*(x)) \\
&\leq \beta^2 \sum_{j=1}^{m}(R_j^2 - R_j)(\tanh^{-1}(u_j^*(x)/\beta))^2 + b_{\epsilon_t} + \|\hat{u}(x) - u^*(x)\|^2 \\
&\leq \epsilon_s + \|\hat{u}(x) - u^*(x)\|^2.
\end{aligned} \tag{5.44}$$

Substituting (5.44) *into* (5.42) *yields*

$$\dot{L}_V \leq - \mathcal{L}(x) - \rho v^*(x)^\top v^*(x) - l_M^2(x) - \rho d_M^2(x) + \epsilon_s + \|\hat{u}(x) - u^*(x)\|^2$$
$$- 2\rho v^{*\top}(x)(\hat{v}(x) - v^*(x))$$
$$= - \mathcal{L}(x) - l_M^2(x) - \rho d_M^2(x) + \epsilon_s - \rho \hat{v}^\top(x)\hat{v}(x) + \|\hat{u}(x) - u^*(x)\|^2$$
$$+ \rho \|\hat{v}(x) - v^*(x)\|^2 .$$

(5.45)

As for $\rho \hat{v}^\top(x)\hat{v}(x)$ *in* (5.45), *according to* (5.38), *we get*

$$\rho \hat{v}^\top(x)\hat{v}(x) = \frac{1}{4\rho}\hat{W}^\top \nabla \Phi(x)h(x)h^\top(x)\nabla \Phi^\top(x)\hat{W}$$
$$= \frac{1}{4\rho}(W^* + \tilde{W})^\top \nabla \Phi(x)h(x)h^\top(x)\nabla \Phi^\top(x)(W^* + \tilde{W})$$
$$= \frac{1}{4\rho}W^{*\top}\mathcal{H}W^* + \frac{1}{4\rho}\tilde{W}^\top \mathcal{H}\tilde{W} + \frac{1}{2\rho}W^{*\top}\mathcal{H}\tilde{W}.$$

(5.46)

where $\mathcal{H} = \nabla \Phi(x)h(x)h^\top(x)\nabla \Phi^\top(x)$.

As for $\rho \|\hat{v}(x) - v^*(x)\|^2$ *in* (5.45), *according to* (5.38), *we get*

$$\rho \|\hat{v}(x) - v^*(x)\|^2 = \rho \left\|\frac{1}{2\rho}h^\top(x)\nabla \Phi(x)\tilde{W}\right\|^2 = \frac{1}{4\rho}\tilde{W}^\top \mathcal{H}\tilde{W}. \qquad (5.47)$$

For simplicity, denote $\mathcal{G}^* = \frac{1}{2\beta}R^{-1}g^\top(x)\nabla \Phi^\top(x)W^*$ *and* $\hat{\mathcal{G}} = \frac{1}{2\beta}R^{-1}g^\top(x)\nabla \Phi^\top(x)\hat{W}$, $\hat{\mathcal{G}} = [\hat{\mathcal{G}}_1, \dots, \hat{\mathcal{G}}_m] \in \mathbb{R}^m$ *with* $\hat{\mathcal{G}}_j \in \mathbb{R}, j = 1, \dots, m$. *Based on* (5.12) *and* (5.37), *the Taylor series of* $\tanh(\mathcal{G}^*)$ *follows*

$$\tanh(\mathcal{G}^*) = \tanh(\hat{\mathcal{G}}) + \frac{\partial \tanh(\hat{\mathcal{G}})}{\partial \hat{\mathcal{G}}}(\mathcal{G}^* - \hat{\mathcal{G}}) + O((\mathcal{G}^* - \hat{\mathcal{G}})^2)$$
$$= \tanh(\hat{\mathcal{G}}) - \frac{1}{2\beta}(I_{m\times m} - \mathcal{D}(\hat{\mathcal{G}}))R^{-1}g^\top(x)\nabla \Phi^\top(x)\tilde{W} + O((\mathcal{G}^* - \hat{\mathcal{G}})^2),$$

(5.48)

where $\mathcal{D}(\hat{\mathcal{G}}) = \text{diag}\tanh^2(\hat{\mathcal{G}}_1), \dots, \tanh^2(\hat{\mathcal{G}}_m))$, $O((\mathcal{G}^* - \hat{\mathcal{G}})^2)$ *is a higher-order term of the Taylor series. By following [106, Lemma 1], the higher-order term is bounded as*

$$\left\|O((\mathcal{G}^* - \hat{\mathcal{G}})^2)\right\| \leq 2\sqrt{m} + \frac{1}{\beta}\left\|R^{-1}\right\|g_M b_{\Phi x}\left\|\tilde{W}\right\|. \qquad (5.49)$$

Using (5.12), (5.37), *and* (5.48), *we get*

$$\hat{u}(x) - u^*(x) = \beta(\tanh(\mathcal{G}^*) - \tanh(\hat{\mathcal{G}})) + \epsilon_u^*$$
$$= -\frac{1}{2}(I_{m\times m} - \mathcal{D}(\hat{\mathcal{G}}))R^{-1}g^\top(x)\nabla \Phi^\top(x)\tilde{W} \qquad (5.50)$$
$$+ \beta O((\mathcal{G}^* - \hat{\mathcal{G}})^2) + \epsilon_u^*.$$

where $\epsilon_u^* = \beta\tanh\left(\frac{1}{2\beta}R^{-1}g^\top(x)(\nabla\Phi^\top(x)W^* + \nabla\epsilon)\right) - \beta\tanh\left(\frac{1}{2\beta}R^{-1}g^\top(x)\nabla\Phi^\top(x)W^*\right)$, *and assuming that it is bounded by* $\|\epsilon_u^*\| \leq b_{\epsilon_u^*}$.

As for $\|\hat{u}(x) - u^*(x)\|^2$ *in* (5.45), *since* $\left\|I_{m\times m} - \mathscr{D}(\hat{\mathscr{G}})\right\| \leq 2$ *[106], combining* (5.49) *with* (5.50), *we get*

$$\|\hat{u}(x) - u^*(x)\|^2 \leq 6\left\|R^{-1}\right\|^2 g_M^2 b_{\Phi x}^2 \left\|\tilde{W}\right\|^2 \\ + 12m\beta^2 + 3b_{\epsilon_u^*}^2 + 12\beta\sqrt{m}\left\|R^{-1}\right\| g_M b_{\Phi x} \left\|\tilde{W}\right\|. \tag{5.51}$$

Substituting (5.46), (5.47), (5.51) *into* (5.45) *yields*

$$\dot{L}_V \leq -\frac{1}{2\rho}W^{*\top}\mathscr{H}\tilde{W} - \mathcal{L}(x) - l_M^2(x) - \rho d_M^2(x) - \frac{1}{4\rho}W^{*\top}\mathscr{H}W^* + \epsilon_s \\ + 6\left\|R^{-1}\right\|^2 g_M^2 b_{\Phi x}^2 \left\|\tilde{W}\right\|^2 + 12m\beta^2 + 3b_{\epsilon_u^*}^2 \\ + 12\beta\sqrt{m}\left\|R^{-1}\right\| g_M b_{\Phi x} \left\|\tilde{W}\right\|. \tag{5.52}$$

As for the second term $\dot{L}_W$, *based on* (5.32) *and* (5.34),

$$\dot{L}_W \leq -\tilde{W}^\top X\tilde{W} + \tilde{W}^\top \epsilon_{er}. \tag{5.53}$$

Finally, as for $\dot{J}$, *substituting* (5.52) *and* (5.53) *into* (5.40), *and based on the fact that* $\|W^*\| \leq b_{W^*}$, $\|\nabla\Phi(x)\| \leq b_{\Phi x}$, $\|h(x)\| \leq h_M$, *we get*

$$\dot{J} \leq -\mathcal{L}(x) - l_M^2(x) - \rho d_M^2(x) - \frac{1}{4\rho}W^{*\top}\mathscr{H}W^* - \tilde{W}^\top X\tilde{W} + M\tilde{W} \\ + 6\left\|R^{-1}\right\|^2 g_M^2 b_{\Phi x}^2 \left\|\tilde{W}\right\|^2 + 12\beta\sqrt{m}\left\|R^{-1}\right\| g_M^2 b_{\Phi x}^2 \left\|\tilde{W}\right\| + 12m\beta^2 + 3b_{\epsilon_u^*}^2 + \epsilon_s \\ \leq -\mathcal{L}(x) - l_M^2(x) - \rho d_M^2(x) - \frac{1}{4\rho}W^{*\top}\mathscr{H}W^* - (\lambda_{\min}(B) - 6\left\|R^{-1}\right\|^2 g_M^2 b_{\Phi x}^2)\left\|\tilde{W}\right\|^2 \\ + 12m\beta^2 + 3b_{\epsilon_u^*}^2 + (12\beta\sqrt{m}\left\|R^{-1}\right\| g_M^2 b_{\Phi x}^2 + b_M)\left\|\tilde{W}\right\| + \epsilon_s \\ = -\mathcal{A} - \mathcal{B}\left\|\tilde{W}\right\|^2 + \mathcal{C}\left\|\tilde{W}\right\| + \mathcal{D}, \tag{5.54}$$

where $M = \epsilon_{er} - \frac{1}{2\rho}W^{*\top}\mathscr{H}$, *and there exists* $b_M = b_{\epsilon_{er}} + \frac{1}{2\rho}b_{\Phi x}^2 h_M^2 b_{W^*} \in \mathbb{R}^+$ *such that* $\|M\| \leq b_M$; $\mathcal{A} = \mathcal{L}(x) + l_M^2(x) + \rho d_M^2(x) + \frac{1}{4\rho}W^{*\top}\mathscr{H}W^*$ *is positive definite*; $\mathcal{B} = \lambda_{\min}(B) - 6\left\|R^{-1}\right\|^2 g_M^2 b_{\Phi x}^2$, $\mathcal{C} = 12\beta\sqrt{m}\left\|R^{-1}\right\| g_M^2 b_{\Phi x}^2 + b_M$ *and* $\mathcal{D} = 12m\beta^2 + 3b_{\epsilon_u^*}^2 + \epsilon_s$.

Let the parameters be chosen such that $\mathcal{B} > 0$. *Since* $\mathcal{A}$ *is positive definite, the above Lyapunov derivative is negative if*

$$\left\|\tilde{W}\right\| > \frac{\mathcal{C}}{2\mathcal{B}} + \sqrt{\frac{\mathcal{C}^2}{4\mathcal{B}^2} + \frac{\mathcal{D}}{\mathcal{B}}}. \tag{5.55}$$

Algorithm 4 Online Prioritized Experience Replay Algorithm

Input: Iteration index: n_r; Buffer size: P; Threshold: ξ.
Output: Experience buffers: $\mathfrak{B}$, $\mathfrak{C}$.
1: **if** $n_r \le P$ **then**
2:　　Record current Y, Θ into $\mathfrak{B}$, $\mathfrak{C}$ respectively.
3: **else**
4:　　**if** $\|W_{n_r} - W_{n_r-1}\| > \xi$ **then**
5:　　　　Record prioritized Y, Θ leading to high $\lambda_{\min}(\mathfrak{B})$.
6:　　**else**
7:　　　　Record current Y, Θ sequentially to update $\mathfrak{B}$,$\mathfrak{C}$.
8:　　**end if**
9: **end if**

Thus, the critic weight learning error converges to the residual set defined as

$$\tilde{\Omega}_{\tilde{W}} = \left\{ \tilde{W} \mid \left\|\tilde{W}\right\| \le \frac{\mathcal{C}}{2\mathcal{B}} + \sqrt{\frac{\mathcal{C}^2}{4\mathcal{B}^2} + \frac{\mathcal{D}}{\mathcal{B}}} \right\}. \tag{5.56}$$

Assumption 5.5 used in Theorem 5.3 is the prerequisite to ensure that $\hat{W}$ converges to W^*. The guaranteed weight convergence enables us to directly apply $\hat{W}$ in (5.32) to construct the approximate optimal control policies (5.37), (5.38). Assumption 5.5 is not restrictive and could be satisfied by the algorithms proposed in the next subsection.

5.2.3   Online and Offline Experience Buffer Construction

To get rich enough experience data to satisfy Assumption 5.5, given the sampling deficiency problem of the sequent way of data usage in existing control-theoretic RL-related works [18, 107, 43], and inspired by the concurrent learning technique developed for system identification [21], here we design both online and offline principled methods to provide the sufficient rich experience data. These recorded informative experience data are then relayed to the weight update law to achieve the required excitation for the guaranteed critic NN weight convergence.

A. Online PER Algorithm

Before the estimated weight converges (lines 4-5), Algorithm 4 chooses the minimum eigenvalue (i.e., $\lambda_{\min}(\mathfrak{B})$) as the priority scheme to filter experience data Y and Θ recorded into the experience buffers $\mathfrak{B}$ and $\mathfrak{C}$,

---

Algorithm 5 Offline Experience Buffer Construction Algorithm

Input: Mesh size : $\delta \in \mathbb{R}^n$, or data point number: $c \in \mathbb{R}^n$;
    $A = [\underline{A}, \overline{A}]$ with $\underline{A}, \overline{A} \in \mathbb{R}^n$; Empty sets: $\mathcal{X} \in \mathbb{R}^{n \times d}$.
Output: $\mathcal{F}; \mathcal{G}; \mathcal{H}; \mathcal{K}; \mathcal{R}; P$.
  1: Sampling: $\mathcal{X}$; $P = \prod_{i=1}^{n}(\overline{A}_i - \underline{A}_i)/\delta_i$, or $\prod_{i=1}^{n} c_i$.
  2: Data collection: $\mathcal{F} \leftarrow \nabla\Phi^\top(\mathcal{X})f(\mathcal{X}); \mathcal{R} \leftarrow r_d(\mathcal{X})$;
    $\mathcal{G} \leftarrow \nabla\Phi^\top(\mathcal{X})g(\mathcal{X}); \mathcal{H} \leftarrow \nabla\Phi^\top(\mathcal{X})h(\mathcal{X}); \mathcal{K} \leftarrow \mathcal{L}(\mathcal{X})$

---

respectively. Here the prioritized criterion is different from ones used in existing PER algorithms [83]. We prefer experience data accompanied with a larger $\lambda_{\min}(\mathfrak{B})$ given the facts that: a) a nonzero $\lambda_{\min}(\mathfrak{B})$ ensures that $rank(\mathfrak{B}) = N$ in Assumption 5.5 holds [21, 52], i.e., the convergence of $\hat{W}$ to W^* is guaranteed; b) according to (5.35) and (5.36), a larger $\lambda_{\min}(\mathfrak{B})$ leads to a faster weight convergence rate and a smaller residual set. Although efficient, the priority scheme $\lambda_{\min}(\mathfrak{B})$ accompanies with additional computation loads. Thus, once the convergence is achieved (lines 6-7), i.e., we have obtained sufficient excitation, we alternate to a low-cost mode where recent data are sequentially recorded. This kind of cyclic replacement way of data usage enjoys robustness to a dynamic environment since collected real-time data could reflect environmental changes in time. Unlike standard methods that first construct a huge experience buffer and then sample partial data [24] to reduce computation loads and relieve hardware requirements, we directly build experience buffers with a limited buffer size P here, and all of the recorded experience data are replayed to the critic NN for the online weight learning. The buffer size P is a hyper-parameter that requires careful tuning. In order to satisfy Assumption 5.5, P is selected such that $P \geq N$ holds.

B. Offline Experience Buffer Construction Algorithm

Algorithm 5 aims to construct experience buffers $\mathcal{F}, \mathcal{G}, \mathcal{H}, \mathcal{K}, \mathcal{R} \in \mathbb{R}^{N \times P}$ full with offline recorded experience data, which are then used to support the online weight learning. For simplicity, here the offline experience data are generated from pre-simulation within the given operation region A. Specifically, for i-th dimension of an allowable operation region $A_i \in \mathbb{R}$, we sample data isometrically with a defined mesh size $\delta_i \in \mathbb{R}^+$, or a prior given number $c_i \in \mathbb{N}^+$. The resulting sampling state space is denoted as $\mathcal{X}$. Note that $\leftarrow$ in Algorithm 5 means that experience data are recorded into corresponding experience buffers. Rather than sampling partial data from the offline constructed experience buffers based on a uniform or a prioritized way [24], we replay all the offline recorded experience data in an average way for the online weight learning. Thereby, the off-policy weight update law (5.32) based on Algorithm 5 is redesigned as

$$\dot{\hat{W}} = -\Gamma k_c Y \tilde{\Theta} - \frac{1}{P} \sum_{l=1}^{P} \Gamma k_e Y_l \tilde{\Theta}_l. \tag{5.57}$$

The implementation of using the offline recorded experience data to support the online learning process enjoys two advantages: the rank condition in Assumption 5.5 is easily satisfied, and the possible influence of data noises is offset by averaging. It is worth mentioning that the mere exploitation of offline recorded data cannot tackle a dynamic environment well. Thus, during the online operation, the offline experience data recorded into experience buffers $\mathcal{F}, \mathcal{G}, \mathcal{H}, \mathcal{K}$, and $\mathcal{R}$ will be sequentially replaced with online counterparts.

CHAPTER 6

Safe Approximate Optimal Control via Filtered RL

6.1 OPTIMAL INCREMENTAL CONTROL

Considering the following continuous time control-affine nonlinear system:

$$\dot{x} = f(x) + g(x)u(x) + d(t), \tag{6.1}$$

where $x \in \mathbb{R}^n$, $u(x) \in \mathbb{R}^m$ are system states and inputs, respectively. $f(x) : \mathbb{R}^n \to \mathbb{R}^n$, $g(x) : \mathbb{R}^n \to \mathbb{R}^{n \times m}$ are continuous and locally Lipschitz drift and input dynamics. $d(t) \in \mathbb{R}^n$ represents a bounded time-varying external disturbance. Assume that no knowledge of dynamics (6.1) is available except for the dimensions of system states and inputs.

The main objective is to tackle the robust stabilization problem of the highly uncertain dynamics (6.1) operating in a disturbed environment, which is formulated as follows.

Problem 6.1 *Design a control strategy $u(x)$ such that the system (6.1) perturbed by a bounded disturbance $d(t)$ is stable under input saturation $\mathbb{U}_j = \{u_j \in \mathbb{R} : |u_j| \leq \beta\}, j = 1, \ldots, m$, where $\beta \in \mathbb{R}^+$ is a known saturation bound.*

Remark 6.1 *Although the explicit form of the controlled plant (6.1) is provided here, which is introduced for the analytical purpose and facilitates the controller design as well as the stability analysis in the following sections, our developed control approach relies on neither model parameters nor environmental information.*

DOI: 10.1201/9781003683650-6

The highly uncertain dynamics (6.1) cannot be directly used to design a controller to solve Problem 6.1. Therefore, based on measured input-state data, we use historical data [34, 108] to get an incremental dynamics that are equivalent to (6.1). This formulated incremental dynamics reflects the system response of the controlled plant (6.1) without using explicit model parameters or preceding identification procedures.

Before proceeding, the following assumption is provided to facilitate the formulation of an incremental dynamics.

Assumption 6.1 *[40] The input dynamics* $g = [g_1, g_2, \ldots, g_m]$ *is bounded, and its columns* $g_1, g_2, \ldots, g_m \in \mathbb{R}^n$ *are linearly independent. The function* $g^\dagger = (g^\top g)^{-1} g^\top : \mathbb{R}^n \to \mathbb{R}^{m \times n}$ *is bounded and locally Lipschitz continuous.*

To get the incremental dynamics, we start with introducing a constant matrix $\bar{g} \in \mathbb{R}^{n \times m}$ and multiply $\bar{g}^\dagger$ on the dynamics (6.1),

$$\bar{g}^\dagger \dot{x} = \bar{g}^\dagger f(x) + \bar{g}^\dagger g(x) u(x) + \bar{g}^\dagger d(t) = H(x, \dot{x}) + u(x), \qquad (6.2)$$

where $H(x, \dot{x}) = (\bar{g}^\dagger - g^\dagger(x))\dot{x} + g^\dagger(x) f(x) + g^\dagger(x) d(t) : \mathbb{R}^n \times \mathbb{R}^n \to \mathbb{R}^m$. It is a lump term that embodies all the unknown model knowledge (i.e., $f(x)$, $g(x)$) as well as external disturbances (i.e., $d(t)$).

Then, with a sufficiently high sampling rate, the unknown $H(x, \dot{x})$ in (6.2) could be estimated by time-delayed signals as

$$\hat{H}(x, \dot{x}) = H(x_0, \dot{x}_0) = \bar{g}^\dagger \dot{x}_0 - u_0, \qquad (6.3)$$

where $x_0 = x(t - L)$, $u_0 = u(x(t - L))$. $L \in \mathbb{R}^+$ is the delay time chosen as one or several sampling periods in practical digital implementations. Given that the smallest achievable L in digital devices is the sampling period [37], thus we finally take the delay time L to be the same as the sampling period to get an accurate estimation of $H(x, \dot{x})$ in (6.3). In other words, x_0, u_0 are the values of states and inputs at the previous sampling period.

Finally, substituting (6.3) into (6.2), we get the incremental dynamics as

$$\Delta \dot{x} = \bar{g} \Delta u + \bar{g} \xi, \qquad (6.4)$$

where $\Delta \dot{x} = \dot{x} - \dot{x}_0 \in \mathbb{R}^n$, and $\Delta u = u(x) - u_0 \in \mathbb{R}^m$ are incremental states and control inputs, respectively. $\xi = H(x, \dot{x}) - \hat{H}(x, \dot{x}) \in \mathbb{R}^m$ denotes the estimation error, which is proved to be bounded as given in Lemma 6.1. Here, with a predefined $\bar{g}$, the measured input-state data (i.e., $\dot{x}$, $\dot{x}_0$, u, and u_0) are adopted to reflect the system response in an incremental way without using model or environmental information.

Remark 6.2 *The so-called sufficiently high sampling rate, which is a prerequisite for estimating the unknown $H(x, \dot{x})$ by reusing past measured input-state data, can be chosen as the value that is larger than 30 times the system bandwidth [37, 27]. In this setting, a digital control system can be regarded as a continuous system so that $H(x, \dot{x})$ in (6.2) does not vary significantly during the sampling period. Thus, the estimation error ξ in (6.4) is sufficiently small.*

However, although an equivalent of (6.1) is provided in (6.4) without using explicit knowledge of dynamics, the unknown estimation error ξ hinders us from directly utilizing (6.4) to design controllers. Therefore, a method will be developed to address the estimation error ξ in the next subsection. Before proceeding, here we first provide the theoretical analysis about the boundness property of ξ, which facilitates the method to tackle the estimation error ξ under an optimization framework.

Lemma 6.1 *Given a sufficiently high sampling rate, $\exists \bar{\xi} \in \mathbb{R}^+$, there holds $\|\xi\| \leq \bar{\xi}$.*

Proof 6.1 *Combining (6.2) with (6.3), the estimation error follows*

$$
\begin{aligned}
\xi &= H(x, \dot{x}) - \hat{H}(x, \dot{x}) = H(x, \dot{x}) - H(x_0, \dot{x}_0) \\
&= (\bar{g}^\dagger - g^\dagger(x))(\dot{x} - \dot{x}_0) + (g_0^\dagger - g^\dagger(x))\dot{x}_0 + g^\dagger(x)f(x) - g_0^\dagger f_0 \\
&\quad + g^\dagger(x)d(t) - g_0^\dagger d_0 \\
&= (\bar{g}^\dagger - g^\dagger(x))\Delta\dot{x} + (g_0^\dagger - g^\dagger(x))\dot{x}_0 + g^\dagger(x)(f(x) - f_0) + (g^\dagger(x) - g_0^\dagger)f_0 \\
&\quad + g^\dagger(x)(d(t) - d_0) + (g^\dagger(x) - g_0^\dagger)d_0.
\end{aligned}
\tag{6.5}
$$

Besides, based on the system (6.1), we get

$$
\begin{aligned}
\Delta\dot{x} &= f(x) + g(x)u(x) + d(t) - f_0 - g_0 u_0 - d_0 \\
&= g(x)\Delta u + (g(x) - g_0)u_0 + f(x) - f_0 + d(t) - d_0.
\end{aligned}
\tag{6.6}
$$

Then, substituting (6.6) into (6.5) yields

$$
\begin{aligned}
\xi &= (\bar{g}^\dagger - g^\dagger(x))g(x)\Delta u + (\bar{g}^\dagger - g^\dagger(x))[(g(x) - g_0)u_0 \\
&\quad + f(x) - f_0 + d(t) - d_0] \\
&\quad + (g_0^\dagger - g^\dagger(x))\dot{x}_0 + g^\dagger(x)(f(x) - f_0) + (g^\dagger(x) - g_0^\dagger)f_0 \\
&\quad + g^\dagger(x)(d(t) - d_0) + (g^\dagger(x) - g_0^\dagger)d_0 \tag{6.7} \\
&= (\bar{g}^\dagger g(x) - I_{m\times m})\Delta u + \delta_1, \tag{6.8}
\end{aligned}
$$

where $\delta_1 = \bar{g}^\dagger(g(x) - g_0)u_0 + \bar{g}^\dagger(f(x) - f_0) + \bar{g}^\dagger(d(t) - d_0)$.

For a sufficiently high sampling rate, the gap between successive states is sufficiently small. Thus, it is reasonable to assume that there exists a positive constant $\bar{\delta}_1 \in \mathbb{R}^+$ such that $\|\delta_1\| \leq \bar{\delta}_1$. In addition, the bounded control input u implies that $\|\Delta u\| \leq 2\beta$ holds. By choosing a suitable $\bar{g}$ such that $\|\bar{g}^\dagger g(x) - I_{m \times m}\| \leq c$ establishes, we could get

$$\|\xi\| \leq \left\|\bar{g}^\dagger g(x) - I_{m \times m}\right\| \|\Delta u\| + \|\delta_1\| \leq c \|\Delta u\| + \bar{\delta}_1 \leq 2\beta c + \bar{\delta}_1 = \bar{\xi}.$$

(6.9)

Remark 6.3 *By using the Taylor series expansion based incremental control technique, previous works [114, 115, 116, 3, 87] attempt to provide the incremental dynamics by offering the first-order approximation of $\dot{x}$ in the neighborhood of $[x_0, u_0]$. It follows*

$$\begin{aligned}
\dot{x} &= f(x) + g(x)u(x) \\
&= f_0 + g_0 u_0 + \frac{\partial[f(x) + g(x)u(x)]}{\partial x}\Big|_{x=x_0, u=u_0}(x - x_0) \\
&\quad + \frac{\partial[f(x) + g(x)u(x)]}{\partial u}\Big|_{x=x_0, u=u_0}(u - u_0) + \mathcal{H.O.T.} \\
&\cong \dot{x}_0 + F[x_0, u_0]\Delta x + G[x_0, u_0]\Delta u,
\end{aligned}$$

where $F[x_0, u_0] = [\partial(f(x) + g(x)u(x))/\partial x]|_{x=x_0, u=u_0} \in \mathbb{R}^{n \times n}$ is the system matrix, and $G[x_0, u_0] = [\partial(f(x)+g(x)u(x))/\partial u]|_{x=x_0, u=u_0} \in \mathbb{R}^{n \times m}$ is the control effectiveness matrix. However, a recursive least square method is demanded to search for suitable gain matrices $F[x_0, u_0]$ and $G[x_0, u_0]$ to construct the incremental dynamics [114, 115, 116]. This required online identification of $F[x_0, u_0]$ and $G[x_0, u_0]$ introduces additional computational burden.

To address the unknown estimation error in the incremental dynamics (6.4), here we attempt to investigate the original robust stabilization problem shown as Problem 6.1 from an optimal control perspective, whereby the estimation error could be reflected in the performance index and further be attenuated during the optimization process. This departs from existing related works [34, 108, 114, 115, 116, 3, 87] that directly ignore the influence of the estimation error on the controller performance. Moreover, the effort to solve Problem 6.1 under an optimization framework enables us to take the desired performance indexes regarding state deviations and control energy expenditures into consideration. These considered performance indexes endow the resulting model-free control strategy with guaranteed optimality.

The estimation error ξ in (6.4) is unknown. Thus, the available incremental dynamics to design a controller to solve Problem 6.1 follows

$$\Delta \dot{x} = \bar{g} \Delta u. \tag{6.10}$$

To attenuate the estimation error ξ that is overlooked in (6.10), as well as to optimize the performance of states and control inputs, we consider the cost function of (6.10) as

$$V(x(t)) = \int_t^\infty r(x(\tau), \Delta u(\tau)) \, d\tau, \tag{6.11}$$

where $r(x, \Delta u) = x^\top Q x + \mathcal{W}(u_0 + \Delta u) + \bar{\xi}_o^2 : \mathbb{R}^n \times \mathbb{R}^m \to \mathbb{R}^+$. The common quadratic positive definite term $x^\top Q x$ reflects users' preference for the controller performance concerning state deviations, where $Q \in \mathbb{R}^{n \times n}$ is a positive definite matrix. The nonquadratic positive definite control penalty function $\mathcal{W}(u_0 + \Delta u)$, which relates to the measured u_0 and to be designed Δu, is introduced to enforce the control limit on $u(x)$ based on the bounded tanh function. The explicit form of this part follows [2]

$$\mathcal{W}(u_0 + \Delta u) = 2 \sum_{j=1}^m \int_0^{u_{0_j} + \Delta u_j} \beta \tanh^{-1}(\vartheta_j/\beta) \, d\vartheta_j. \tag{6.12}$$

where $\vartheta_j \in \mathbb{R}^m$. Originally, we could incorporate the quadratic estimation error bound $\bar{\xi}^2$ into $r(x, \Delta u)$ to attenuate the estimation error ξ during the optimization process. However, according to (6.9) of Lemma 6.1, the explicit value of $\bar{\xi}$ is unknown. Thus, we seek for a bounded $\bar{\xi}_o^2$, where $\bar{\xi}_o = \bar{c} \|\Delta u\|$ and $\bar{c} \in \mathbb{R}^+$ is chosen as illustrated in Theorem 6.1, to replace $\bar{\xi}^2$ to accomplish the same goal. It is worth noting that the designed utility function $r(x, \Delta u)$ here enables us to perform the optimization of incremental control inputs.

Remark 6.4 *Note that there exist other options to address the estimation error ξ. For example, by treating the unknown estimation error ξ in (6.4) as a kind of disturbance, we can introduce the widely used disturbance-observer based methods [20] or sliding mode control methods [86] to compensate the estimation error ξ. Comparing to these add-on methods, our strategy enjoys computational simplicity.*

The above settings allow us to formulate an optimal incremental control problem presented as Problem 6.2, whose equivalence to Problem 6.1 will be later proved in Theorem 6.1.

Problem 6.2 *Given Assumption 6.1 and Lemma 6.1, consider the incremental dynamics (6.10), find an incremental control strategy Δu to minimize the cost function defined as (6.11).*

Before proceeding to formally solve Problem 6.2, by following [2, Definition 1] where admissible controls are defined based on (6.1), here we define the set of incremental control inputs that are considered admissible for Problem 6.2. The admissible incremental control given in Definition 6.1 facilitates the following derivation of the closed-form optimal incremental control strategy.

Definition 6.1 (Admissible incremental control) *An incremental control $\Delta\mu(x)$ is defined to be admissible with respect to (6.11) on $\Omega \subseteq \mathbb{R}^n$, denoted by $\Delta\mu(x) \in \Psi(\Omega)$, if $\Delta\mu(x)$ is continuous on Ω, $\Delta\mu(0) = 0$, $\Delta u(x) = \Delta\mu(x)$ stabilizes (6.10) on Ω, and $V(x)$ is finite $\forall x \in \Omega$.*

For any admissible incremental control policies $\Delta u \in \Psi(\Omega)$, using Leibniz's rule [26] to differentiate V in (6.11) yields the following relation

$$0 = r(x, \Delta u) + \nabla V^T \dot{x} = r(x, \Delta u) + \nabla V^T (\Delta \dot{x} + \dot{x}_0)$$
$$= r(x, \Delta u) + \nabla V^T (\bar{g}\Delta u + \dot{x}_0). \tag{6.13}$$

Define the Hamiltonian function as

$$H(x, \Delta u, \nabla V) = r(x, \Delta u) + \nabla V^T (\bar{g}\Delta u + \dot{x}_0). \tag{6.14}$$

Let $V^*(x)$ be the optimal cost function defined as

$$V^*(x) = \min_{\Delta u \in \Psi(\Omega)} \int_t^\infty r(x(\tau), \Delta u(\tau)) \, d\tau. \tag{6.15}$$

Combining with (6.14), $V^*(x)$ satisfies the HJB equation

$$0 = \min_{\Delta u \in \Psi(\Omega)} [H(x, \Delta u, \nabla V^*)]. \tag{6.16}$$

Assume that the minimum on the right side of (6.16) exists and is unique [98]. By using the stationary optimality condition, i.e., $\partial H(x, \Delta u, \nabla V^*)/\partial \Delta u = 0$, we get the closed-form optimal incremental control strategy as

$$\Delta u^* = -\beta \tanh\left(\frac{1}{2\beta} \bar{g}^T \nabla V^*\right) - u_0. \tag{6.17}$$

Then, we could construct the corresponding optimal control strategy as

$$u^* = u_0 + \Delta u^* = -\beta \tanh\left(\frac{1}{2\beta}\bar{g}^{\top}\nabla V^*\right). \tag{6.18}$$

Departing from traditional approximate dynamic programming (ADP) related works [98, 40] where the total optimal control input u^* is directly designed, here we first get the theoretically derived incremental optimal control strategy Δu^* in (6.17), and then construct u^* based on the measured u_0 and the designed Δu^*. This difference lies in that Problem 6.2 is formulated based on the incremental dynamics (6.10) that relates to incremental states and control inputs.

Remark 6.5 *Alternatively, we could use $\mathcal{W}(\Delta u) = 2\sum_{j=1}^{m}\int_{0}^{\Delta u_j}\alpha$ $\tanh^{-1}(\vartheta_j/\alpha)\,d\vartheta_j$. This enforces the constraint satisfaction of the incremental control inputs, which is denoted as $-\alpha \leq \Delta u_j \leq \alpha$, $\alpha \in \mathbb{R}^+$, $j = 1,\ldots,m$. By following the aforementioned derivation processes (6.15)-(6.18), the corresponding optimal incremental control follows $\Delta u^* = -\alpha \tanh(\frac{1}{2\alpha}\bar{g}^{\top}\nabla V^*)$. Then, the resulting optimal control is $u^* = u_0 + \Delta u^*$. However, in this case, the control limit on $u(x)$ cannot be addressed. Given that input saturation is common in real life and violations of it might lead to serious consequences, we prefer to incorporate (6.12) into $r(x, \Delta u)$ to enforce the control limit on $u(x)$.*

To get Δu^* (6.17) and u^* (6.18), ∇V^* remains to be determined. We defer the explicit method to acquire ∇V^* in Section 6.2, and focus now on the equivalence proof to show that after solving Problem 6.2, the resulting u^* (6.18) constructed from the designed Δu^* (6.17) is the robust stabilization solution to Problem 6.1.

Theorem 6.1 *Given Assumption 6.1 and Lemma 6.1, consider the system described by (6.1), if there exists a scalar $\bar{c} \in \mathbb{R}^+$ such that*

$$\bar{\xi} < \bar{c}\,\|\Delta u\|, \tag{6.19}$$

the system (6.1) is robustly stabilized by the optimal control strategy (6.18) with the optimal incremental control strategy (6.17).

Proof 6.2 *Given that $V^*(x = 0) = 0$, and $V^* > 0$ for $\forall x \neq 0$, V^* defined in (6.15) could serve as a Lyapunov function candidate for the stability proof.*

Taking time derivative of V^* along the incremental dynamics (6.4), which is an equivalent of the original dynamics (6.1), we get

$$\dot{V}^* = \nabla V^{*\top}(\Delta \dot{x} + \dot{x}_0) = \nabla V^{*\top}(\bar{g}\Delta u^* + \bar{g}\xi + \dot{x}_0) = \nabla V^{*\top}(\bar{g}\Delta u^* + \dot{x}_0) + \nabla V^{*\top}\bar{g}\xi.$$
(6.20)

According to (6.16) and (6.17), the following equations hold:

$$\nabla V^{*\top}(\bar{g}\Delta u^* + \dot{x}_0) = -x^\top Q x - \mathcal{W}(u_0 + \Delta u^*) - \bar{\xi}_o^2, \ \nabla V^{*\top}\bar{g} = -2\beta \tanh^{-1}\left(\frac{u_0 + \Delta u^*}{\beta}\right).$$
(6.21)

Substituting (6.21) into (6.20) reads

$$\dot{V}^* = -x^\top Q x - \mathcal{W}(u_0 + \Delta u^*) - \bar{\xi}_o^2 - 2\beta \tanh^{-1}\left(\frac{u_0 + \Delta u^*}{\beta}\right)\xi. \quad (6.22)$$

As for $\mathcal{W}(u_0 + \Delta u^*)$ in (6.22), based on the explicit form in (6.12) and by setting $\varsigma_j = \tanh^{-1}(\vartheta_j/\beta)$, it follows

$$
\begin{aligned}
\mathcal{W}(u_0 + \Delta u^*) &= 2\beta \sum_{j=1}^{m} \int_0^{u_{0_j} + \Delta u_j^*} \tanh^{-1}(\vartheta_j/\beta)\, d\vartheta_j \\
&= 2\beta^2 \sum_{j=1}^{m} \int_0^{\tanh^{-1}\left(\frac{u_{0_j} + \Delta u_j^*}{\beta}\right)} \varsigma_j(1 - \tanh^2(\varsigma_j))\, d\varsigma_j \\
&= \beta^2 \sum_{j=1}^{m} \left(\tanh^{-1}\left(\frac{u_{0_j} + \Delta u_j^*}{\beta}\right)\right)^2 - \epsilon_u, \quad (6.23)
\end{aligned}
$$

where $\epsilon_u = 2\beta^2 \sum_{j=1}^{m} \int_0^{\tanh^{-1}\left(\frac{u_{0_j} + \Delta u_j^*}{\beta}\right)} \varsigma_j \tanh^2(\varsigma_j)\, d\varsigma_j$. Based on the integral mean-value theorem, there exists a series of $\theta_j \in [0, \tanh^{-1}\left(\frac{u_{0_j} + \Delta u_j^*}{\beta}\right)], j = 1, \ldots, m$, such that

$$\epsilon_u = 2\beta^2 \sum_{j=1}^{m} \tanh^{-1}\left(\frac{u_{0_j} + \Delta u_j^*}{\beta}\right) \theta_j \tanh^2(\theta_j). \quad (6.24)$$

Based on (6.21) and the fact $0 \le \tanh^2(\theta_j) \le 1$, it follows

$$
\begin{aligned}
\epsilon_u &\le 2\beta^2 \sum_{j=1}^{m} \left(\frac{u_{0_j} + \Delta u_j^*}{\beta}\right)\theta_j \\
&\le 2\beta^2 \sum_{j=1}^{m} \left(\tanh^{-1}\left(\frac{u_{0_j} + \Delta u_j^*}{\beta}\right)\right)^2 \\
&= \frac{1}{2}\nabla V^{*\top}\bar{g}\bar{g}^\top \nabla V^*. \quad (6.25)
\end{aligned}
$$

The definition of admissible incremental control in Definition 6.1 implies that V^* is finite. Additionally, there exists $b_{\nabla V^*} \in \mathbb{R}^+$ such that $\|\nabla V^*\| \leq b_{\nabla V^*}$. Thus, we could rewrite (6.25) as

$$\epsilon_u \leq b_{\epsilon_u} = \frac{1}{2} \|\bar{g}\|^2 b_{\nabla V^*}^2. \tag{6.26}$$

Then, substituting (6.23), (6.26) into (6.22) yields

$$\dot{V}^* \leq -x^\top Q x - (\bar{\xi}_o^2 - \|\xi\|^2) - [\beta \tanh^{-1}\left(\frac{u_0 + \Delta u^*}{\beta}\right) + \xi]^2 + b_{\epsilon_u}. \tag{6.27}$$

By choosing $\bar{\xi}_o = \bar{c} \|\Delta u\|$, and $\bar{c}$ is chosen to satisfy $\bar{c} \|\Delta u\| > \bar{\xi}$, where $\bar{\xi}$ is defined in (6.9), the following inequality holds

$$\dot{V}^* \leq -x^\top Q x + b_{\epsilon_u}. \tag{6.28}$$

Thus, $\dot{V}^* < 0$ holds if $-\lambda_{\min}(Q) \|x\|^2 + b_{\epsilon_u} < 0$. Finally, it concludes that states converge to the residual set

$$\Omega_x = \{x | \|x\| \leq \sqrt{b_{\epsilon_u}/\lambda_{\min}(Q)}\}. \tag{6.29}$$

The aforementioned proof means that based on the optimal cost function (6.15), the derived optimal incremental control policy (6.17) of the system (6.10) robustly stabilizes the system (6.4). Given the equivalence between (6.1) and (6.4), thus the optimal control input (6.18), which is constructed from the designed (6.17), robustly stabilizes the system (6.1).

We have proved in Theorem 6.1 that the optimal incremental control problem clarified in Problem 6.2 is equivalent to the robust stabilization problem shown as Problem 6.1. Thus, to stabilize the highly uncertain dynamics (6.1) operating in a disturbed environment, the remaining part devotes to solving Problem 6.2.

6.2 APPROXIMATE OPTIMAL INCREMENTAL POLICY

To solve Problem 6.2, this section seeks for the approximate solution to the value function of the HJB equation (6.16) that is hard to solve directly. Departing from common ADP-related works [98, 40] using an actor-critic learning structure, we use a single critic learning structure here, which decreases the computational burden and simplifies the theoretical analysis.

6.2.1 Approximated Value Function

Based on the Weierstrass high-order approximation theorem [25], for $x \in \Omega$ with $\Omega \subset \mathbb{R}^n$ being a compact set, the optimal value function is approximated as [98]

$$V^*(x) = W^{*\top}\Phi(x) + \epsilon(x), \qquad (6.30)$$

where $W^* \in \mathbb{R}^N$ is a weighting matrix, $\Phi(x) : \mathbb{R}^n \to \mathbb{R}^N$ represents the activation function, and $\epsilon(x) \in \mathbb{R}$ denotes the approximation error. The partial derivative of $V^*(x)$ follows

$$\nabla V^*(x) = \nabla \Phi^\top(x)W^* + \nabla \epsilon(x), \qquad (6.31)$$

where $\nabla \Phi(x) \in \mathbb{R}^{N \times n}$, $\nabla \epsilon(x) \in \mathbb{R}^n$. As $N \to \infty$, both $\epsilon(x)$ and $\nabla \epsilon(x)$ converge to zero uniformly. Without loss of generality, the following assumption is given, which is common in ADP-related works.

Assumption 6.2 *[98] There exist constants $b_\epsilon, b_{\epsilon x}, b_\Phi, b_{\Phi x} \in \mathbb{R}^+$ such that $\|\epsilon(x)\| \leq b_\epsilon$, $\|\nabla \epsilon(x)\| \leq b_{\epsilon x}$, $\|\Phi(x)\| \leq b_\Phi$, and $\|\nabla \Phi(x)\| \leq b_{\Phi x}$.*

Considering a fixed incremental control input Δu, inserting (6.31) into (6.16) yields

$$W^{*\top}\nabla\Phi(\bar{g}\Delta u + \dot{x}_0) + r(x, \Delta u) = \epsilon_h, \qquad (6.32)$$

where the residual error follows $\epsilon_h = -\nabla\epsilon^\top(\bar{g}\Delta u + \dot{x}_0) \in \mathbb{R}$. Assume that there exists $b_{\epsilon_h} \in \mathbb{R}^+$ such that $\|\epsilon_h\| \leq b_{\epsilon_h}$. Focusing on the NN parameterized (6.32), we rewrite it into the following LIP form

$$\Theta = -W^{*\top}Y + \epsilon_h, \qquad (6.33)$$

where $\Theta = r(x, \Delta u) \in \mathbb{R}$, and $Y = \nabla\Phi(\bar{g}\Delta u + \dot{x}_0) \in \mathbb{R}^N$. Given that Θ and Y could be obtained from real-time data, this formulated LIP form enables the learning of W^* to be equivalent to a parameter identification problem of an LIP system from the perspective of adaptive control. The above applied transformation allows us to directly use our developed off-policy weight update law in the previous chapter to solve Problem 6.2.

6.2.2 NN Weight Update Law

Following our previous results, the critic NN weight updates as

$$\dot{\hat{W}} = -\Gamma k_c Y \tilde{\Theta} - \sum_{l=1}^{P} \Gamma k_e Y_l \tilde{\Theta}_l, \qquad (6.34)$$

to get the approximate solution to the HJB function (6.16).

The guaranteed weight convergence of $\hat{W}$ to W^* permits us to directly use the estimated critic NN weight $\hat{W}$ to construct the approximate optimal incremental control strategy. Therefore, based on the optimal incremental control strategy in (6.17), the approximate optimal incremental control strategy follows

$$\Delta\hat{u} = -\beta\tanh\left(\frac{1}{2\beta}\bar{g}^{\mathsf{T}}\nabla\Phi^{\mathsf{T}}\hat{W}\right) - u_0. \tag{6.35}$$

Accordingly, the approximate optimal control strategy applied at the plant (6.1) follows

$$\hat{u} = u_0 + \Delta\hat{u} = -\beta\tanh\left(\frac{1}{2\beta}\bar{g}^{\mathsf{T}}\nabla\Phi^{\mathsf{T}}\hat{W}\right). \tag{6.36}$$

Remark 6.6 *From a practical perspective, our designed model-free approximate optimal incremental control strategy (6.35) only requires one manually tuned constant matrix $\bar{g}$. This feature of our method decreases the required parameter tuning efforts compared to existing identification based methods to fulfil model-free control strategies [12, 60, 109, 97, 14, 92], where multiple hyperparameters or gains need to be tuned.*

Based on the off-policy weight update law (6.34), and the approximate optimal incremental control strategy (6.35) mentioned above, we provide the main conclusions as follows.

Theorem 6.2 *Consider the incremental dynamics (6.10), the off-policy weight update law of the critic NN in (6.34), and the approximate optimal incremental control policy (6.35). Given Assumptions 6.1-6.2, for a sufficiently large N, the approximate optimal incremental control policy (6.35) stabilizes the incremental dynamics (6.10), and the critic NN weight learning error $\tilde{W}$ is uniformly ultimately bounded.*

Proof 6.3 *Consider the following candidate Lyapunov function*

$$J = V^*(x) + \frac{1}{2}\tilde{W}^{\mathsf{T}}\Gamma^{-1}\tilde{W}. \tag{6.37}$$

Denoting $\dot{L}_V = \dot{V}^*(x)$ and $\dot{L}_W = \tilde{W}^{\mathsf{T}}\Gamma^{-1}\dot{\tilde{W}}$, the time derivative of (6.37) reads

$$\dot{J} = \dot{L}_V + \dot{L}_W. \tag{6.38}$$

The first term $\dot{L}_V$ follows

$$\dot{L}_V = \nabla V^{*\top}(\bar{g}\Delta\hat{u} + \bar{g}\xi + \dot{x}_0) = \nabla V^{*\top}(\bar{g}\Delta u^* + \dot{x}_0) + \nabla V^{*\top}\bar{g}\xi + \nabla V^{*\top}\bar{g}(\Delta\hat{u} - \Delta u^*).$$
(6.39)

Then, substituting (6.21) into (6.39) gets

$$\dot{L}_V = -x^\top Q x - \mathcal{W}(u_0 + \Delta u^*) - \bar{\xi}_o^2 - 2\beta\tanh^{-1}\left(\frac{u_0 + \Delta u^*}{\beta}\right)\xi$$
$$- 2\beta\tanh^{-1}\left(\frac{u_0 + \Delta u^*}{\beta}\right)(\Delta\hat{u} - \Delta u^*).$$
(6.40)

According to (6.23)-(6.25), (6.40) follows

$$\dot{L}_V \leq -x^\top Q x - (\bar{\xi}_o^2 - \|\xi\|^2) - [\beta\tanh^{-1}\left(\frac{u_0 + \Delta u^*}{\beta}\right) + \xi]^2 + \frac{1}{2}\nabla V^{*\top}\bar{g}\bar{g}^\top\nabla V^*$$
$$- 2\beta\tanh^{-1}\left(\frac{u_0 + \Delta u^*}{\beta}\right)(\Delta\hat{u} - \Delta u^*).$$
(6.41)

The term $-2\beta\tanh^{-1}\left(\frac{u_0+\Delta u^*}{\beta}\right)(\Delta\hat{u} - \Delta u^*)$ in (6.41) follows

$$-2\beta\tanh^{-1}\left(\frac{u_0 + \Delta u^*}{\beta}\right)(\Delta\hat{u} - \Delta u^*) \leq \beta^2\left\|\tanh^{-1}\left(\frac{u_0 + \Delta u^*}{\beta}\right)\right\|^2 + \|\Delta\hat{u} - \Delta u^*\|^2.$$
(6.42)

By using (6.17), (6.31), and the mean-value theorem, the optimal incremental control is rewritten as

$$\Delta u^* = -\beta\tanh\left(\frac{1}{2\beta}\bar{g}^\top\nabla\Phi^\top W^*\right) - \epsilon_{\Delta u^*} - u_0,$$
(6.43)

where $\epsilon_{\Delta u^*} = \frac{1}{2}(\mathbf{1} - \tanh^2(\eta))\bar{g}^\top\nabla\epsilon$, and $\eta \in \mathbb{R}^m$ is chosen between $\frac{1}{2\beta}\bar{g}^\top\nabla\Phi^\top W^*$ and $\frac{1}{2\beta}\bar{g}^\top\nabla V^*$, $\mathbf{1} = [1,\ldots,1]^\top \in \mathbb{R}^m$. According to $\|\nabla\epsilon\| \leq b_{\epsilon x}$ in Assumption 6.2, $\|\epsilon_{\Delta u^*}\| \leq \frac{1}{2}\|\bar{g}\|\,b_{\epsilon x}$ holds. Then, combining (6.35) with (6.43), we get

$$\Delta\hat{u} - \Delta u^* = \beta(\tanh\left(\frac{1}{2\beta}\bar{g}^\top\nabla\Phi^\top W^*\right) - \tanh\left(\frac{1}{2\beta}\bar{g}^\top\nabla\Phi^\top\hat{W}\right) + \epsilon_{\Delta u^*}.$$
(6.44)

Denoting $\mathscr{G}^* = \frac{1}{2\beta}\bar{g}^\top\nabla\Phi^\top W^*$, and $\hat{\mathscr{G}} = \frac{1}{2\beta}\bar{g}^\top\nabla\Phi^\top\hat{W}$, where $\hat{\mathscr{G}} = [\hat{\mathscr{G}}_1,\ldots,\hat{\mathscr{G}}_m] \in \mathbb{R}^m$ with $\hat{\mathscr{G}}_j \in \mathbb{R}, j = 1,\ldots,m$. Based on (6.17) and (6.35), the Taylor series of $\tanh(\mathscr{G}^*)$ follows

$$\tanh(\mathscr{G}^*) = \tanh\left(\hat{\mathscr{G}}\right) + \frac{\partial\tanh\left(\hat{\mathscr{G}}\right)}{\partial\hat{\mathscr{G}}}(\mathscr{G}^* - \hat{\mathscr{G}}) + O((\mathscr{G}^* - \hat{\mathscr{G}})^2)$$
$$= \tanh\left(\hat{\mathscr{G}}\right) - \frac{1}{2\beta}(I_{m\times m} - \mathscr{D}(\hat{\mathscr{G}}))\bar{g}^\top\nabla\Phi^\top\tilde{W} + O((\mathscr{G}^* - \hat{\mathscr{G}})^2),$$
(6.45)

where $\mathscr{D}(\hat{\mathscr{G}}) = \mathrm{diag}(\tanh^2(\hat{\mathscr{G}}_1), \ldots, \tanh^2(\hat{\mathscr{G}}_m))$, and $O(\mathscr{G}^* - \hat{\mathscr{G}})^2$ is a higher-order term of the Taylor series. By following [106, Lemma 1], this higher-order term is bounded as

$$\left\| O((\mathscr{G}^* - \hat{\mathscr{G}})^2) \right\| \le 2\sqrt{m} + \frac{1}{\beta} \|\bar{g}\| \, b_{\Phi x} \left\| \tilde{W} \right\|. \tag{6.46}$$

Based on (6.45), we rewrite (6.44) as

$$\Delta\hat{u} - \Delta u^* = \beta(\tanh(\mathscr{G}^*) - \tanh\left(\hat{\mathscr{G}}\right)) + \epsilon_{\Delta u^*}$$

$$= -\frac{1}{2}(I_{m \times m} - \mathscr{D}(\hat{\mathscr{G}}))\bar{g}\nabla\Phi^{\top}\tilde{W} + \beta O((\mathscr{G}^* - \hat{\mathscr{G}})^2) + \epsilon_{\Delta u^*}. \tag{6.47}$$

According to [106], $\left\| I_{m \times m} - \mathscr{D}(\hat{\mathscr{G}}) \right\| \le 2$ holds. Then, combining (6.46) with (6.47), $\|\Delta\hat{u} - \Delta u^*\|^2$ in (6.42) follows

$$\|\Delta\hat{u} - \Delta u^*\|^2 \le 3\beta^2 \left\| O((\mathscr{G}^* - \hat{\mathscr{G}})^2) \right\|^2 + 3 \|\epsilon_{\Delta u^*}\|^2 + 3 \left\| -\frac{1}{2}(I_{m \times m} - \mathscr{D}(\hat{\mathscr{G}}))\bar{g}^{\top}\nabla\Phi^{\top}\tilde{W} \right\|^2$$

$$\le 6 \|\bar{g}\|^2 b_{\Phi x}^2 \left\| \tilde{W} \right\|^2 + 12m\beta^2 + \frac{3}{4} \|\bar{g}\|^2 b_{\epsilon x}^2 + 12\beta\sqrt{m} \|\bar{g}\| \, b_{\Phi x} \left\| \tilde{W} \right\|. \tag{6.48}$$

Based on (6.21), (6.31), Assumption 6.2, and the fact that $\|W^*\| \le b_{W^*}$, $\left\| \tanh^{-1}((u_0 + \Delta u^*)/\beta) \right\|^2$ in (6.42) follows

$$\left\| \tanh^{-1}\left(\frac{u_0 + \Delta u^*}{\beta}\right) \right\|^2 = \left\| \frac{1}{4\beta^2} \nabla V^{*\top} \bar{g}\bar{g}^{\top} \nabla V^* \right\|$$

$$\le \frac{1}{4\beta^2} \|\bar{g}\|^2 b_{\Phi x}^2 b_{W^*}^2 + \frac{1}{4\beta^2} b_{\epsilon x}^2 \|\bar{g}\|^2 + \frac{1}{2\beta^2} \|\bar{g}\|^2 b_{\Phi x} b_{\epsilon x} b_{W^*}. \tag{6.49}$$

Using (6.48) and (6.49), (6.42) reads

$$-2\beta \tanh^{-1}\left(\frac{u_0 + \Delta u^*}{\beta}\right)(\Delta\hat{u} - \Delta u^*) \le \frac{1}{4} \|\bar{g}\|^2 b_{\Phi x}^2 b_{W^*}^2 + \frac{1}{4} b_{\epsilon x}^2 \|\bar{g}\|^2$$

$$+ \frac{1}{2} \|\bar{g}\|^2 b_{\Phi x} b_{\epsilon x} b_{W^*} + 12m\beta^2 + 6 \|\bar{g}\|^2 b_{\Phi x}^2 \left\| \tilde{W} \right\|^2$$

$$+ \frac{3}{4} \|\bar{g}\|^2 b_{\epsilon x}^2 + 12\beta\sqrt{m} \|\bar{g}\| \, b_{\Phi x} \left\| \tilde{W} \right\|.$$

Substituting (6.50) into (6.41), finally the first term $\dot{L}_V$ follows

$$\dot{L}_V \le -x^{\top}Qx - (\bar{\xi}_o^2 - \xi^{\top}\xi) - [\beta \tanh^{-1}\left(\frac{u_0 + \Delta u^*}{\beta}\right) + \xi]^2 + \frac{3}{4} \|\bar{g}\|^2 b_{\Phi x}^2 b_{W^*}^2$$

$$+ \frac{3}{4} b_{\epsilon x}^2 \|\bar{g}\|^2 + \frac{3}{2} \|\bar{g}\|^2 b_{\Phi x} b_{\epsilon x} b_{W^*} + 6 \|\bar{g}\|^2 b_{\Phi x}^2 \left\| \tilde{W} \right\|^2 + 12m\beta^2$$

$$+ \frac{3}{4} \|\bar{g}\|^2 b_{\epsilon x}^2 + 12\beta\sqrt{m} \|\bar{g}\| \, b_{\Phi x} \left\| \tilde{W} \right\|. \tag{6.50}$$

As for the second term $\dot{L}_W$, based on (6.34) , it follows

$$\dot{L}_W \leq -\tilde{W}^\top B \tilde{W} + \tilde{W}^\top \epsilon_{\tilde{W}}. \tag{6.51}$$

Finally, as for $\dot{J}$, substituting (6.50) and (6.51) into (6.38), we get

$$\dot{J} \leq -\mathcal{A} - \mathcal{B} \left\| \tilde{W} \right\|^2 + \mathcal{C} \left\| \tilde{W} \right\| + \mathcal{D}, \tag{6.52}$$

where $\mathcal{A} = x^\top Q x + (\bar{\xi}_o^2 - \xi^\top \xi) + [\beta \tanh^{-1} \left(\frac{u_0 + \Delta u^*}{\beta} \right) + \xi]^2$, $\mathcal{B} = \lambda_{\min}(B) - 6 \left\| \bar{g} \right\|^2 b_{\Phi x}^2$, $\mathcal{C} = 12\beta\sqrt{m} \left\| \bar{g} \right\| b_{\Phi x} + \bar{\epsilon}_{\tilde{W}}$, and $\mathcal{D} = \frac{3}{4} \left\| \bar{g} \right\|^2 b_{\Phi x}^2 b_{W^*}^2 + \frac{3}{2} b_{\epsilon x}^2 \left\| \bar{g} \right\|^2 + \frac{3}{2} \left\| \bar{g} \right\|^2 b_{\Phi x} b_{\epsilon x} b_{W^*} + 12m\beta^2$. Let the parameters be chosen such that $\mathcal{B} > 0$. Since $\mathcal{A}$ is positive definite, the above Lyapunov derivative (6.52) is negative if $\left\| \tilde{W} \right\| > \frac{\mathcal{C}}{2\mathcal{B}} + \sqrt{\frac{\mathcal{C}^2}{4\mathcal{B}^2} + \frac{\mathcal{D}}{\mathcal{B}}}$. Thus, the critic weight learning error converges to the residual set $\bar{\Omega}_{\tilde{W}} = \left\{ \tilde{W} \mid \left\| \tilde{W} \right\| \leq \frac{\mathcal{C}}{2\mathcal{B}} + \sqrt{\frac{\mathcal{C}^2}{4\mathcal{B}^2} + \frac{\mathcal{D}}{\mathcal{B}}} \right\}$.

6.3 SAFETY FILTER IMPLEMENTATION

Under a satisfying framework [91], this section introduces a safety filter to correct the learned approximate optimal incremental control policy (6.36) via a minimally invasive way to ensure safe operation.

The safety filter is implemented as a CBF based QP formulation:

$$
\begin{aligned}
u_s &= \arg \min_{u_s} \left\| u_s - \hat{u} \right\| \\
\text{s.t.}\ \ &\ddot{h}_j + \alpha_{1_j} \dot{h}_j + \alpha_{2_j} h_j \geq 0,\ j = 1, 2, \ldots
\end{aligned}
\tag{6.53}
$$

where $u_s \in \mathbb{R}^n$ is the corrected safe control input; h_j is the j-th HO-CBF characterizing the j-th unsafe region, which is prior-given or learned via the method developed in Chapter 3; α_{1_j}, $\alpha_{2_j} \in \mathbb{R}^+$ are chosen using our optimized ACS method to guarantee that the utilized h_j is a valid HO-CBF. The barrier certified approximate optimal control policy from (6.53) are used for future data collection to support the value function learning process illustrated in Section 6.2.

The presented QP (6.53) implies the potential conflict between safety and performance. For practical applications, safety should be prioritized over performance. Therefore, the solution is to safely achieve a performance that is as close as possible to the desired performance. This is achieved by (6.53), wherein the relaxation of strict optimality allows us to introduce safety considerations into our method. This add-on safety filter method enjoys flexibility toward multiple tasks and environments, although the promising theoretical optimality guarantee is lost. This is satisfying for practical applications.

CHAPTER 7

Time-Delayed Data Informed RL for Optimal Tracking Control

7.1 NONLINEAR OPTIMAL TRACKING CONTROL

The investigated plant (unknown dynamics) is described by the E-L equation:

$$M(q)\ddot{q} + N(q, \dot{q}) + F(\dot{q}) = \tau, \tag{7.1}$$

where $M(q) : \mathbb{R}^n \to \mathbb{R}^{n \times n}$ is the symmetric positive definite inertia matrix; $N(q, \dot{q}) = C(q, \dot{q})\dot{q} + G(q) : \mathbb{R}^n \times \mathbb{R}^n \to \mathbb{R}^n$, $C(q, \dot{q}) : \mathbb{R}^n \times \mathbb{R}^n \to \mathbb{R}^{n \times n}$ is the matrix of centrifugal and Coriolis terms, $G(q) : \mathbb{R}^n \to \mathbb{R}^n$ represents the gravitational terms; $F(\dot{q}) : \mathbb{R}^n \to \mathbb{R}^n$ denotes the viscous friction; q, $\dot{q}$, $\ddot{q} \in \mathbb{R}^n$ are the vectors of angles, velocities, and accelerations, respectively; and $\tau \in \mathbb{R}^n$ represents the input torque vector. Note that the mathematical model (7.1) is provided here for later theoretical analysis. The explicit values of $M(q)$, $C(q, \dot{q})$, $G(q)$, and $F(\dot{q})$ are unavailable to practitioners.

The objective is to design a model-free tracking control strategy τ to enable the plant (7.1) to track a bounded and smooth reference signal $x_d = [q_d^\top, \dot{q}_d^\top]^\top \in \mathbb{R}^{2n}$ while minimizing a predefined performance function. The considered high-dimensional and highly uncertain controlled plant (7.1) provides difficulty in solving the nonlinear optimal tracking control problem (OTCP) mentioned above.

DOI: 10.1201/9781003683650-7

7.2 INCREMENTAL SUBSYSTEM

We utilize the decoupled control technique and time-delayed signals [33, 37] to develop model-free incremental subsystems. The formulated incremental subsystems are equivalent to the dynamics (7.1), but no explicit model information is required. Specifically, the decoupled control technique is utilized to divide the high-dimensional system into multiple low-dimensional subsystems. Then, time-delayed data is used to estimate the unknown dynamics as well as the decoupled control-related coupling terms. Here, the constructed incremental subsystems serve as the basis to design the model-free tracking control strategy in Section 7.3 and allow us to address the scalability problem of the value function approximation in Section 7.4.1.

The high-dimensional system (7.1) can be decoupled into multiple subsystems, wherein the i-th subsystem reads

$$M_{ii}\ddot{q}_i + \sum_{j=1,j\neq i}^{n} M_{ij}\ddot{q}_j + N_i + F_i = \tau_i, \quad i = 1, 2, \cdots, n. \qquad (7.2)$$

Let $x_i = [x_{i_1}, x_{i_2}]^\top = [q_i, \dot{q}_i]^\top \in \mathbb{R}^2$, and $u_i = \tau_i \in \mathbb{R}$. We rewrite (7.2) as

$$\dot{x}_{i_1} = x_{i_2}, \qquad (7.3a)$$
$$\dot{x}_{i_2} = f_i + g_i u_i, \qquad (7.3b)$$

where $f_i = -(\sum_{j=1,j\neq i}^{n} M_{ij}\ddot{q}_j + N_i + F_i)/M_{ii} \in \mathbb{R}$, and $g_i = 1/M_{ii} \in \mathbb{R}$ are unknown. Clearly, f_i and g_i are upper bounded since $M(q)$, $N(q,\dot{q})$, and $F(\dot{q})$ in (7.1) are upper bounded [57]. Throughout this article, each subsystem is assumed to be controllable.

The unknown functions f_i and g_i hinder us to directly design tracking controllers based on the subsystem (7.3). Departing from common methods that identify the unknown f_i, g_i explicitly through a tedious identification process [113, 39, 66, 61, 9, 73], we exploit time-delayed signals to estimate the unknown model knowledge. To achieve time delay estimation, we first introduce a predetermined constant $\bar{g}_i \in \mathbb{R}^+$ and multiply $\bar{g}_i^{-1}$ on (7.3b),

$$\bar{g}_i^{-1}\dot{x}_{i_2} = h_i + u_i, \qquad (7.4)$$

where $h_i = (\bar{g}_i^{-1} - g_i^{-1})\dot{x}_{i_2} + g_i^{-1}f_i \in \mathbb{R}$ is a lumped term that embodies the unknown model knowledge f_i, g_i of (7.3b).

Then, with a sufficiently high sampling rate, by utilizing time-delayed signals [34, 108, 59], the unknown h_i in (7.4) could be estimated as

$$\hat{h}_i = h_{i,0} = \bar{g}_i^{-1} \dot{x}_{i_{2,0}} - u_{i,0}, \tag{7.5}$$

where $u_{i,0} = u_i(t - L)$, $\dot{x}_{i_{2,0}} = \dot{x}_{i_2}(t - L)$. We choose the delay time $L \in \mathbb{R}^+$ as the sampling period (the smallest achievable value of L in practical implementations) to achieve an accurate estimation of h_i [59].

Substituting (7.5) into (7.4), we get

$$\dot{x}_{i_2} = \dot{x}_{i_{2,0}} + \bar{g}_i(\Delta u_i + \xi_i), \tag{7.6}$$

where $\Delta u_i = u_i - u_{i,0} \in \mathbb{R}$ is the incremental control input; $\xi_i = h_i - \hat{h}_i \in \mathbb{R}$ denotes the estimation error that is proved to be bounded in Lemma 7.1 of Section 7.3.

Combining (7.3) with (7.6), we finally obtain the i-th incremental subsystem dynamics

$$\dot{x}_{i_1} = x_{i_2}, \tag{7.7a}$$

$$\dot{x}_{i_2} = \dot{x}_{i_{2,0}} + \bar{g}_i(\Delta u_i + \xi_i), \tag{7.7b}$$

which is an equivalent of the original i-th subsystem (7.3) but without using explicit model information. The guideline to select the required suitable $\bar{g}_i$ to construct the i-th incremental subsystem (7.7) is provided in Remark 7.1. Here the time-delayed data ($\dot{x}_{i_{2,0}}$ and $u_{i,0}$ in particular) informs the value function learning process clarified in Section 7.4 about one model-free representation (7.7) of the original controlled plant (7.1). Thereby, we could achieve model-free control and also have a mathematical form of dynamics to conduct rigorous theoretical analysis using rich analysis tools from the control field.

Remark 7.1 *According to [57], it is reasonable to assume that* $\underline{m}_i \leq M_{ii} \leq \overline{m}_i$, *where* $\underline{m}_i, \overline{m}_i \in \mathbb{R}^+$. *According to (7.3),* $g_i = \frac{1}{M_{ii}}$. *Thus,* $\frac{1}{\overline{m}_i} \leq g_i \leq \frac{1}{\underline{m}_i}$ *holds. To achieve* $\left\| 1 - g_i(k)\bar{g}_i^{-1} \right\| < 1$ *required in (7.31),* $\bar{g}_i > \frac{1}{2}g_i$ *needs to be satisfied. Therefore, we could choose* $\bar{g}_i > \frac{1}{2\underline{m}_i}$. *The prior knowledge of* M_{ii} *provides designers with hints to choose a suitable* $\bar{g}_i$.

This section has decoupled the original n-D (7.1) into n equivalent 2-D incremental subsystems (7.7). Accordingly, we transform the OTCP of (7.1) into n sub-OTCPs regarding (7.7). The following section will present our developed tracking control scheme by focusing on the sub-OTCP of (7.7).

Remark 7.2 *The decoupled control technique facilitates real-time control for a high-dimensional system by distributing the computation load into multiple processors. However, the utilized decoupled control technique presents a challenge of getting the value of the coupling terms, which is usually addressed by add-on tools such as (RBF) NNs [111, 67] that accompany with additional parameter tuning efforts and computational loads. Unlike these works, the time-delayed signals, which are initially used to achieve model-free control in a low-cost and easily implemented way (only a constant $\bar{g}_i$ to be debugged), enjoys an additional benefit that compensates the coupling terms in (7.2).*

Remark 7.3 *The required state derivative information (i.e., $\dot{x}_{i2,0}$) to construct the incremental subsystem dynamics (7.7) may not be directly measurable. In practice, the unmeasurable state derivative is usually obtained via numerical differentiation [33, 19]. Alternatively, the state derivative could be estimated by the robust exact differentiator [53], or derivative estimator [41, 11], which is beyond the scope of this chapter.*

7.3 TRACKING CONTROL SCHEME

This section details our proposed tracking control scheme to solve the sub-OTCP of (7.7). The incremental control input to be designed follows

$$\Delta u_i = \Delta u_{i_f} + \Delta u_{i_b}, \tag{7.8}$$

where the incremental dynamic inversion based $\Delta u_{i_f} \in \mathbb{R}$ serves to transform the time-varying sub-OTCPs into equivalent time-invariant sub-robust optimal regulation control problems (sub-RORCPs) in Section 7.3.1; and $\Delta u_{i_b} \in \mathbb{R}$ is the incremental control policy to optimally drive the tracking error to zero in Section 7.3.2. The detailed procedures to design Δu_{i_f} and Δu_{i_b} are detailly clarified in Section 7.3.1 and Section 7.3.2, respectively.

7.3.1 Incremental Error Subsystem

This subsection formulates the i-th incremental error subsystem via the properly chosen Δu_{i_f}. The formulated incremental error subsystem converts the sub-OTCP regarding (7.7) into its sub-RORCP, and facilitates the development of the optimal incremental control policy in Section 7.3.2. The detailed procedures to design Δu_{i_f} and to generate the incremental error subsystem are as follows.

Let $e_i = [e_{i_1}, e_{i_2}]^\top \in \mathbb{R}^2$, where $e_{i_1} = x_{i_1} - q_{d_i} \in \mathbb{R}$ and $e_{i_2} = x_{i_2} - \dot{q}_{d_i} \in \mathbb{R}$. Combining with (7.7b) and (7.8) yields

$$\dot{e}_{i_2} = \dot{x}_{i_{2,0}} + \bar{g}_i(\Delta u_{i_f} + \Delta u_{i_b} + \xi_i) - \ddot{x}_{r_i}. \tag{7.9}$$

Designing the required Δu_{i_f} in (7.9) as

$$\Delta u_{i_f} = \bar{g}_i^{-1}(\ddot{x}_{r_i} - \dot{x}_{i_{2,0}} - k_{i_1} e_{i_1} - k_{i_2} e_{i_2}), \tag{7.10}$$

and substituting (7.10) into (7.9), we get

$$\dot{e}_{i_2} = -k_{i_1} e_{i_1} - k_{i_2} e_{i_2} + \bar{g}_i \Delta u_{i_b} + \bar{g}_i \xi_i, \tag{7.11}$$

where $k_{i_1}, k_{i_2} \in \mathbb{R}^+$. Recall that $\dot{e}_{i_1} = e_{i_2}$. Then, combining with (7.11), we obtain the i-th incremental error subsystem

$$\dot{e}_i = A_i e_i + B_i \Delta u_{i_b} + B_i \xi_i, \tag{7.12}$$

where $A_i = \begin{bmatrix} 0 & 1 \\ -k_{i_1} & -k_{i_2} \end{bmatrix} \in \mathbb{R}^{2 \times 2}$, and $B_i = \begin{bmatrix} 0 \\ \bar{g}_i \end{bmatrix} \in \mathbb{R}^2$. The sub-OTCP of (7.7) illustrated in Section 7.2 aims to drive the values of e_i to zero in an optimal manner. This is equivalent to the sub-RORCP of the incremental error subsystem (7.12) given the unknown ξ_i. In other words, this subsection transforms the sub-OTCP of (7.7) into the sub-RORCP regarding (7.12) by designing Δu_{i_f} in the form of (7.10).

Remark 7.4 *The developed Δu_{i_f} (7.10) here acts as a supplementary control input to the Δu_{i_b} designed in Section 7.3.2. In particular, the utilized Δu_{i_f} generates an incremental error subsystem (7.12). Then, we train Δu_{i_b} in Section 7.3.2 based on the incremental error subsystem formulated in this subsection. This practice departs from most of existing approximate dynamic programming related works for the OTCP [40, 109, 72], wherein the tracking control strategies are trained on one specific reference trajectory dynamics. Thus, the flexibility of our developed tracking control scheme against varying desired trajectories is improved without directly using reference signals during the learning process.*

7.3.2 Optimal Incremental Policy

This subsection develops an optimal incremental control policy to solve the sub-RORCP of (7.12), i.e., robustly stabilizing the tracking error to zero in an optimal manner. Departing from common solutions to OTCPs [40, 109, 72], we additionally introduce an estimation error-related term into the value function such that the influence of the estimation error on the controller performance is lessened under an optimization framework.

Given ξ_i in (7.12) is unknown, thus the available incremental error subsystem for later analysis follows

$$\dot{e}_i = A_i e_i + B_i \Delta u_{i_b}. \tag{7.13}$$

To stabilize (7.13) in an optimal manner, the value function is considered as

$$V_i(t) = \int_t^\infty r_i(e_i(\nu), \Delta u_{i_b}(\nu)) \, d\nu, \tag{7.14}$$

where $r_i(e_i, \Delta u_{i_b}) = e_i^\top Q_i e_i + W_i(\Delta u_{i_b}) + \bar{\xi}_{oi}^2$. The quadratic term $e_i^\top Q_i e_i$, where $Q_i \in \mathbb{R}^{2 \times 2}$ is a positive definite matrix, is introduced to improve tracking precision. The input penalty function $W_i(\Delta u_{i_b})$ follows

$$\mathcal{W}_i(\Delta u_{i_b}) = 2 \int_0^{\Delta u_{i_b}} \beta \tanh^{-1}(\vartheta/\beta) \, d\vartheta, \tag{7.15}$$

which is utilized to punish and enforce the optimal incremental control input as $\|\Delta u_{i_b}\| \leq \beta \in \mathbb{R}^+$. The limited Δu_{i_b} is beneficial since a severe interruption might lead to an abrupt change of Δu_{i_b}, which might destabilize the learning process introduced in Section 7.4. The utilized estimation error-related term $\bar{\xi}_{oi}^2$ in $r_i(e_i, \Delta u_{i_b})$ allows designers to attenuate the estimation error during the optimization process. The explicit form of $\bar{\xi}_{oi}$ follows $\bar{\xi}_{oi} = \bar{c}_i \|\Delta u_{i_b}\|$, where $\bar{c}_i \in \mathbb{R}^+$. The rationality of designing $\bar{\xi}_{oi}$ in the above form and the requirement for an appropriate $\bar{c}_i$ are provided in Theorem 7.1.

For $\Delta u_{i_b} \in \Psi$, where Ψ is the set of admissible incremental control policies [59, Definition 1], the associated optimal value function follows

$$V_i^* = \min_{\Delta u_{i_b} \in \Psi} \int_t^\infty r_i(e_i(\nu), \Delta u_{i_b}(\nu)) \, d\nu. \tag{7.16}$$

Define the Hamiltonian function as

$$H_i(e_i, \Delta u_{i_b}, \nabla V_i) = r(e_i, \Delta u_{i_b}) + \nabla V_i^T (A_i e_i + B_i \Delta u_{i_b}), \tag{7.17}$$

where $\nabla(\cdot) = \partial(\cdot)/\partial e_i$. Then, V_i^* satisfies the HJB equation

$$0 = \min_{\Delta u_{i_b} \in \Psi} [H_i(e_i, \Delta u_{i_b}, \nabla V_i^*)]. \tag{7.18}$$

Assume that the minimum of (7.16) exists and is unique [59, 98]. By using the stationary optimality condition on the HJB equation (7.18), we gain an analytical-form optimal incremental control strategy:

$$\Delta u_{i_b}^* = -\beta \tanh\left(\frac{1}{2\beta} B_i^\top \nabla V_i^*\right). \tag{7.19}$$

To obtain $\Delta u_{i_b}^*$, we need to solve the HJB equation (7.18) to determine the value of ∇V_i^*, which is detailly clarified in Section 7.4. In the following part of this subsection, based on the estimation error bound given in Lemma 7.1, we prove in Theorem 7.1 that the optimal incremental control policy $\Delta u_{i_b}^*$ (7.19) regarding (7.13) is the solution to the sub-RORCP of (7.12).

Lemma 7.1 *Given a sufficiently high sampling rate, $\exists \bar{\xi}_i \in \mathbb{R}^+$, there holds $\|\xi_i\| \le \bar{\xi}_i$.*

Proof 7.1 *Combining (7.4) with (7.5), the estimation error for the i-th subsystem (7.7) follows*

$$\xi_i = h_i - \hat{h}_i = h_i - h_{i,0}$$
$$= (\bar{g}_i^{-1} - g_i^{-1})\Delta \dot{x}_{i_2} + (g_{i,0}^{-1} - g_i^{-1})\dot{x}_{i_2,0} + g^{-1}(f_i - f_{i,0}) + (g_i^{-1} - g_{i,0}^{-1})f_{i,0}, \tag{7.20}$$

where $\Delta \dot{x}_{i_2} = \dot{x}_{i_2} - \dot{x}_{i_2,0}$. Combining with (7.3b), (7.7b) and (7.8), $\Delta \dot{x}_{i_2}$ follows

$$\Delta \dot{x}_{i_2} = f_i + g_i u_i - f_{i,0} - g_{i,0} u_{i,0} = g_i \Delta u_i + (g_i - g_{i,0})u_{i,0} + f_i - f_{i,0}$$
$$= g_i(\Delta u_{i_f} + \Delta u_{i_b}) + (g_i - g_{i,0})u_{i,0} + f_i - f_{i,0}. \tag{7.21}$$

Substituting (7.21) into (7.20), we get

$$\xi_i = (g_i \bar{g}_i^{-1} - 1)\Delta u_{i_f} + (g_i \bar{g}_i^{-1} - 1)\Delta u_{i_b} + \delta_{1i}, \tag{7.22}$$

where $\delta_{1i} = \bar{g}_i^{-1}(g_i - g_{i,0})u_0 + \bar{g}_i^{-1}(f_i - f_{i,0})$.

Denoting $\mu_i = \ddot{x}_{r_i} - k_{i_1} e_{i_1} - k_{i_2} e_{i_2} \in \mathbb{R}$. According to (7.5) and (7.10), Δu_{i_f} in (7.22) follows

$$\Delta u_{i_f} = \bar{g}_i^{-1}(\mu_i - \bar{g}_i h_{i,0} - \bar{g}_i u_{i,0})$$
$$= \bar{g}_i^{-1}\mu_i - (\bar{g}_i^{-1} - g_{i,0}^{-1})\dot{x}_{i_2,0} - g_{i,0}^{-1}f_{i,0} - u_{i,0}$$
$$= \bar{g}_i^{-1}\mu_i - (\bar{g}_i^{-1} - g_{i,0}^{-1})(f_{i,0} + g_{i,0}u_{i,0}) - g_{i,0}^{-1}f_{i,0} - u_{i,0}$$
$$= \bar{g}_i^{-1}\mu_i - \bar{g}_i^{-1}(f_{i,0} + g_{i,0}u_{i,0}) = \bar{g}_i^{-1}(\mu_i - \mu_{i,0}) - \bar{g}_i^{-1}(\dot{x}_{i_2,0} - \mu_{i,0}), \tag{7.23}$$

where $\mu_{i,0} = \ddot{x}_{r_{i},0} - k_{i_1} e_{i_1,0} - k_{i_2} e_{i_2,0}$. Besides, combining (7.7b) with (7.8), we get

$$\begin{aligned}
\dot{x}_{i_2} &= \dot{x}_{i_2,0} + \bar{g}_i(\Delta u_{i_f} + \Delta u_{i_b}) + \bar{g}_i \xi_i \\
&= \dot{x}_{i_2,0} + \bar{g}_i \bar{g}_i^{-1}(\mu_i - \dot{x}_{i_2,0}) + \bar{g}_i \Delta u_{i_b} + \bar{g}_i \xi_i \\
&= \mu_i + \bar{g}_i \Delta u_{i_b} + \bar{g}_i \xi_i.
\end{aligned} \tag{7.24}$$

Based on the result shown in (7.24), we get

$$\xi_i = \bar{g}_i^{-1}(\dot{x}_{i_2} - \mu_i - \bar{g}_i \Delta u_{i_b}). \tag{7.25}$$

Accordingly, the following equation establishes

$$\xi_{i,0} = \bar{g}_i^{-1}(\dot{x}_{i_2,0} - \mu_{i,0} - \bar{g}_i \Delta u_{i_b,0}). \tag{7.26}$$

Based on the result given in (7.26), (7.23) is rewritten as

$$\begin{aligned}
\Delta u_{i_f} &= \bar{g}_i^{-1}(\mu_i - \mu_{i,0}) - \bar{g}_i^{-1}(\dot{x}_{i_2,0} - \mu_{i,0} - \bar{g}_i \Delta u_{i_b,0}) - \Delta u_{i_b,0} \\
&= \bar{g}_i^{-1}(\mu_i - \mu_{i,0}) - \xi_{i,0} - \Delta u_{i_b,0}.
\end{aligned} \tag{7.27}$$

Substituting (7.27) into (7.22) yields

$$\xi_i = (1 - g_i \bar{g}_i^{-1})\xi_{i,0} + (1 - g_i \bar{g}_i^{-1})\bar{g}_i^{-1}(\mu_{i,0} - \mu_i) + (1 - g_i \bar{g}_i^{-1})(\Delta u_{i_b,0} - \Delta u_{i_b}) + \delta_{1i}. \tag{7.28}$$

In discrete-time domain, (7.28) could be represented as

$$\xi_i(k) = (1 - g_i(k)\bar{g}_i^{-1})\xi_i(k-1) + (1 - g_i(k)\bar{g}_i^{-1})\Delta \tilde{u}_{i_b} + \delta_{1i} + \delta_{2i}, \tag{7.29}$$

where $\Delta \tilde{u}_{i_b} = \Delta u_{i_b}(k-1) - \Delta u_{i_b}(k)$, $\delta_{2i} = (1 - g_i(k)\bar{g}_i^{-1})\bar{g}_i^{-1}(\mu_i(k-1) - \mu_i(k))$.

The constrained input $\|\Delta u_{i_b}(k)\| \leq \beta$ implies that the following equation holds

$$\|\Delta \tilde{u}_{i_b}\| \leq \|\Delta u_{i_b}(k-1)\| + \|\Delta u_{i_b}(k)\| \leq 2\beta. \tag{7.30}$$

We choose the value of $\bar{g}_i$ to meet $\left\|1 - g_i(k)\bar{g}_i^{-1}\right\| \leq \iota_i < 1$, where $\iota_i \in \mathbb{R}^+$. Under a sufficiently high sampling rate, it is reasonable to assume that there exists $\bar{\delta}_{1i}, \bar{\delta}_{2i} \in \mathbb{R}^+$ such that $\|\delta_{1i}\| \leq \bar{\delta}_{1i}$, and $\|\delta_{2i}\| \leq \iota_i \bar{\delta}_{2i}$.

Then, the following equations hold:

$$\|\xi_i(k)\| \leq \iota_i \|\xi_i(k-1)\| + \iota_i \|\Delta \tilde{u}_{i_b}\| + \bar{\delta}_{1i} + \iota_i \bar{\delta}_{2i}$$
$$\leq \iota_i^2 \|\xi_i(k-2)\| + (\iota_i^2 + \iota_i) \|\Delta \tilde{u}_{i_b}\| + (\iota_i + 1)(\bar{\delta}_{1i} + \iota_i \bar{\delta}_{2i})$$
$$\leq \cdots$$
$$\leq \iota_i^k \|\xi_i(0)\| + \frac{\bar{\delta}_{1i} + \iota_i \bar{\delta}_{2i}}{1 - \iota_i} + \frac{\iota_i \|\Delta \tilde{u}_{i_b}\|}{1 - \iota_i} \qquad (7.31)$$
$$\leq \iota_i^k \|\xi_i(0)\| + \frac{\bar{\delta}_{1i} + \iota_i \bar{\delta}_{2i}}{1 - \iota_i} + \frac{2\iota_i \beta}{1 - \iota_i} = \bar{\xi}_i.$$

As $k \to \infty$, $\bar{\xi}_i \to \frac{\bar{\delta}_{1i} + \iota_i \bar{\delta}_{2i}}{1 - \iota_i} + \frac{2\iota_i \beta}{1 - \iota_i}$.

Theorem 7.1 *Consider the system (7.12) with a sufficiently high sampling rate, if there exists a scalar* $\bar{c}_i \in \mathbb{R}^+$ *such that the following inequality is satisfied*

$$\bar{\xi}_i < \bar{c}_i \|\Delta u_{i_b}\|, \qquad (7.32)$$

the optimal incremental control policy (7.19) regulates the tracking error to a small neighborhood around zero while minimizing the value function (7.14).

Proof 7.2 V_i^* *is a positive definite function, i.e.,* $V_i^*(e_i) \geq 0$ *and iff* $e_i = 0$, $V_i^*(e_i) = 0$. *Thus,* V_i^* *could serve as a candidate Lyapunov function. Taking time derivative of* V_i^* *along the i-th incremental error subsystem (7.12) yields*

$$\dot{V}_i^* = \nabla V_i^*(A_i e_i + B_i \Delta u_{i_b}^*) + \nabla V_i^* B_i \xi_i. \qquad (7.33)$$

According to (7.17) and (7.18), the following equations establish

$$\nabla V_i^*(A_i e_i + B_i \Delta u_{i_b}^*) = -e_i^\top Q_i e_i - W_i(\Delta u_{i_b}^*) - \bar{\xi}_{oi}^2$$
$$\nabla V_i^* B_i = -2\beta \tanh^{-1}(\Delta u_{i_b}^*/\beta). \qquad (7.34)$$

Substituting (7.34) into (7.33) yields

$$\dot{V}_i^* = -e_i^\top Q_i e_i - W_i(\Delta u_{i_b}^*) - \bar{\xi}_{oi}^2 - 2\beta \tanh^{-1}(\Delta u_{i_b}^*/\beta)\xi_i. \qquad (7.35)$$

As for the $W_i(\Delta u_{i_b}^*)$ *in (7.35), according to our previous result [59, Theorem 1], it follows that*

$$W_i(\Delta u_{i_b}^*) = \beta^2 \sum_{j=1}^{m} \left(\tanh^{-1}(\Delta u_{i_b}^*/\beta)\right)^2 - \epsilon_{u_i}, \qquad (7.36)$$

where $\epsilon_{u_i} \le \frac{1}{2}\bar{g}_i^2 \nabla V_i^{*\top} \nabla V_i^*$. Given that there exists $b_{\nabla V_i^*} \in \mathbb{R}^+$ such that $\|\nabla V_i^*\| \le b_{\nabla V_i^*}$. Thus, we could rewrite the bound of ϵ_{u_i} as $\epsilon_{u_i} \le b_{\epsilon u i} \le \frac{1}{2}\bar{g}_i^2 b_{\nabla V_i^*}^2$.

Then, substituting (7.36) into (7.35), we get

$$\dot{V}_i^* = -e_i^\top Q_i e_i - [\beta \tanh^{-1}(\Delta u_{i_b}^*/\beta) + \xi_i]^2 - (\bar{\xi}_{oi}^2 - \xi_i^\top \xi_i) + b_{\epsilon u i}. \tag{7.37}$$

We choose $\bar{\xi}_{oi} = \bar{c}_i \|\Delta u_{i_b}\|$, and $\bar{c}_i$ is picked to satisfy $\bar{c}_i \|\Delta u_{i_b}\| > \bar{\xi}_i$, where $\bar{\xi}_i$ is defined in (7.31). Then, the following equation holds

$$\dot{V}_i^* \le -e_i^\top Q_i e_i + b_{\epsilon u i}. \tag{7.38}$$

Thus, if $-\lambda_{\min}(Q_i) \|e_i\|^2 + b_{\epsilon u i} < 0$, $\dot{V}_i^* < 0$ holds. Here $\lambda_{\min}(\cdot)$ denotes the minimum eigenvalues of a symmetric real matrix. Finally, it concludes that states of the i-th incremental error subsystem (7.12) converges to the residual set

$$\Omega_{e_i} = \{e_i| \|e_i\| \le \sqrt{b_{\epsilon u i}/\lambda_{\min}(Q_i)}\}. \tag{7.39}$$

Theorem 7.1 implies that the optimal incremental control policy $\Delta u_{i_b}^*$ (7.19) robustly stabilize (7.12). It has been clarified in Section 7.3.1 that the sub-RORCP of (7.12) equals to the sub-OTCP of (7.7) based on our designed Δu_{i_f} (7.10). Thus, the designed $\Delta u_{i_b}^*$ and Δu_{i_f} solve the sub-OTCP of (7.7) together.

7.4 APPROXIMATE SOLUTIONS

This section uses a parallel critic learning structure to seek for the approximate solutions to the value functions of the HJB equations (7.18) of n incremental error subsystems (7.12). By reinvestigating the online NN weight learning process from a parameter identification perspective, we develop a simple yet efficient off-policy critic NN weight update law with guaranteed weight convergence by exploiting real-time and experience data together.

7.4.1 Value Function Approximation

For $e_i \in \Omega$, where $\Omega \subset \mathbb{R}^2$ is a compact set, the continuous optimal value function (7.16) is approximated by an critic agent as [98]

$$V_i^* = W_i^{*\top} \Phi_i(e_i) + \epsilon_i(e_i), \tag{7.40}$$

where $W_i^* \in \mathbb{R}^{N_i}$ is the critic NN weight, $\Phi_i(e_i) : \mathbb{R}^2 \to \mathbb{R}^{N_i}$ represents the activation function, and $\epsilon_i(e_i) \in \mathbb{R}$ denotes the approximation error.

Remark 7.5 *The utilized decoupled control technique in Section 7.2 solves the curse of complexity problem in (7.40). In particular, the constructed critic NN (7.40) relies on the error $e_i \in \mathbb{R}^2$ of the incremental error subsystem (7.12). The 2-D e_i allows us to construct a low-dimensional $\Phi_i(e_i)$ (easy to choose) to approximate its associated V_i^* regardless of the value of the system dimension n. For example, the 4-D activation functions $\Phi_i(e_i)$ in a fixed structure are chosen for subsystems of a 3-DoF robot manipulator , and 6-DoF quadrotor. Otherwise, for a global approximation, i.e., $V^* = W^{*\top}\Phi(e) + \epsilon(e)$ with the tracking error $e = x - x_d \in \mathbb{R}^{2n}$, the dimension of $\Phi(e)$ increases exponentially as n increases.*

To facilitate the later theoretical analysis, an assumption that is common in related works is provided here.

Assumption 7.1 *[98] There exist constants $b_{\epsilon_i}, b_{\epsilon_{ei}}, b_{\epsilon_{hi}}, b_{\Phi_i}, b_{\Phi_{ei}} \in \mathbb{R}^+$ such that $\|\epsilon_i(e_i)\| \leq b_{\epsilon_i}$, $\|\nabla\epsilon_i(e_i)\| \leq b_{\epsilon_{ei}}$, $\|\epsilon_{hi}\| \leq b_{\epsilon_{hi}}$, $\|\Phi_i(e_i)\| \leq b_{\Phi_i}$, and $\|\nabla\Phi_i(e_i)\| \leq b_{\Phi_{ei}}$.*

Given a fixed incremental control input Δu_{i_b}, combining (7.18) with (7.40) yields

$$W_i^{*\top}\nabla\Phi_i(A_ie_i + B_i\Delta u_{i_b}) + r_i(e_i, \Delta u_{i_b}) = \epsilon_{h_i}, \qquad (7.41)$$

where the residual error $\epsilon_{h_i} = -\nabla\epsilon_i^\top(A_ie_i + B_i\Delta u_{i_b}) \in \mathbb{R}$. The NN parameterized (7.41) is able to be written into an LIP form as

$$\Theta_i = -W_i^{*\top}Y_i + \epsilon_{h_i}, \qquad (7.42)$$

where $\Theta_i = r_i(e_i, \Delta u_{i_b}) \in \mathbb{R}$, and $Y_i = \nabla\Phi_i(A_ie_i + B_i\Delta u_{i_b}) \in \mathbb{R}^{N_i}$. The values of Θ_i and Y_i are both available to practitioners given the measurable e_i and Δu_{i_b}. This formulated LIP form (7.42) enables the learning of W_i^* to be equivalent to a parameter identification problem of an LIP system, which facilitates the development of an efficient weight update law in the subsequent subsection.

7.4.2 Critic NN Weight Update Law

An approximation of (7.42) follows

$$\hat{\Theta}_i = -\hat{W}_i^\top Y_i, \qquad (7.43)$$

where $\hat{W}_i \in \mathbb{R}^{N_i}$, $\hat{\Theta}_i \in \mathbb{R}$ are estimates of W_i^* and Θ_i, respectively. To enable $\hat{W}_i$ converge to W_i^*, we design an off-policy critic NN weight update law for each subsystem as

$$\dot{\hat{W}}_i = -\Gamma_i k_{t_i} Y_i \tilde{\Theta}_i - \sum_{l=1}^{P_i} \Gamma_i k_{e_i} Y_{i_l} \tilde{\Theta}_{i_l}, \qquad (7.44)$$

to update the critic NN weight $\hat{W}_i$ in a parallel way to minimize $E_i = \frac{1}{2}\tilde{\Theta}_i^\top \tilde{\Theta}_i$, where $\tilde{\Theta}_i = \Theta_i - \hat{\Theta}_i \in \mathbb{R}$. Here $\Gamma_i \in \mathbb{R}^{N_i \times N_i}$ is a constant positive definite gain matrix; $k_{t_i}, k_{e_i} \in \mathbb{R}^+$ are used to trade-off the contribution of real-time and experience data to the online NN weight learning process; $P_i \in \mathbb{R}^+$ is the number of the utilized recorded experience data.

To guarantee the weight convergence of (7.44), as proved in Theorem 7.2, the exploited experience data should be sufficient rich to satisfy the rank condition in Assumption 7.2. This assumption could be satisfied by sequentially reusing experience data in practice [59].

Assumption 7.2 *Given an experience buffer $\mathfrak{B}_i = [Y_{i_1}, ..., Y_{i_{P_i}}] \in \mathbb{R}^{N_i \times P_i}$, there holds $rank(\mathfrak{B}_i) = N_i$.*

Theorem 7.2 *Given Assumption 7.2, the NN weight learning error $\tilde{W}_i$ converges to a small neighborhood around zero.*

Proof 7.3 *The proof is similar to Chapter 4. Thus, it is omitted here for simplicity.*

The guaranteed weight convergence of $\hat{W}_i$ to W_i^* in Theorem 7.2 permits us to use a computation-simple single critic NN learning structure for each subsystem, where the estimated critic NN weight $\hat{W}_i$ is directly used to construct the approximate optimal incremental control strategy:

$$\Delta \hat{u}_{i_b} = -\beta \tanh\left(\frac{1}{2\beta} B_i^\top \nabla \Phi_i^\top \hat{W}_i\right). \qquad (7.45)$$

Finally, combining with (7.8), (7.10), and (7.45), we get the overall control input applied at the i-th subsystem (7.3)

$$\hat{u}_i = u_{i,0} + \Delta u_{i_f} + \Delta \hat{u}_{i_b}. \qquad (7.46)$$

Based on the theoretical analysis mentioned above, we provide the main conclusions in the following theorem.

Theorem 7.3 *Given Assumptions 7.1–7.2, for a sufficiently large N_i, the off-policy critic NN weight update law (7.44), and the approximate optimal incremental control policy (7.45) guarantee the tracking error and the NN weight learning error uniformly ultimately bounded.*

Proof 7.4 *Consider the candidate Lyapunov function for the i-th incremental error subsystem (7.12) as*

$$L_i = V_i^* + \frac{1}{2}\tilde{W}_i^\top \Gamma_i^{-1}\tilde{W}_i. \tag{7.47}$$

By denoting $L_{i_1} = V_i^*$, its derivative follows

$$\dot{L}_{i_1} = \nabla V_i^{*\top}(A_i e_i + B_i \Delta \hat{u}_{i_b} + B_i \xi_i)$$
$$= \nabla V_i^{*\top}(A_i e_i + B_i \Delta u_{i_b}^*) + \nabla V_i^{*\top} B_i \xi_i + \nabla V_i^{*\top} B_i(\Delta \hat{u}_{i_b} - \Delta u_{i_b}^*). \tag{7.48}$$

Substituting (7.34) into (7.48) reads

$$\dot{L}_{i_1} = -e_i^\top Q_i e_i - \mathcal{W}(\Delta u_{i_b}^*) - \bar{\xi}_{oi}^2 - 2\beta \tanh^{-1}(\Delta u_{i_b}^*/\beta)\,\xi_i$$
$$- 2\beta \tanh^{-1}(\Delta u_{i_b}^*/\beta)\,(\Delta \hat{u}_{i_b} - \Delta u_{i_b}^*). \tag{7.49}$$

Combining with (7.36) and (7.37), (7.49) follows

$$\dot{L}_{i_1} \leq -e_i^\top Q_i e_i - (\bar{\xi}_{oi}^2 - \|\xi_i\|^2) - \left[\beta \tanh^{-1}(\Delta u_{i_b}^*/\beta) + \xi_i\right]^2$$
$$+ \frac{1}{2}\nabla V_i^{*\top} B_i B_i^\top \nabla V_i^* - 2\beta \tanh^{-1}(\Delta u_{i_b}^*/\beta)\,(\Delta \hat{u}_{i_b} - \Delta u_{i_b}^*). \tag{7.50}$$

The term $-2\beta \tanh^{-1}(\Delta u_{i_b}^*/\beta)\,(\Delta \hat{u}_{i_b} - \Delta u_{i_b}^*)$ in (7.50) follows

$$-2\beta \tanh^{-1}(\Delta u_{i_b}^*/\beta)\,(\Delta \hat{u}_{i_b} - \Delta u_{i_b}^*) \leq \beta^2 \left\|\tanh^{-1}(\Delta u_{i_b}^*/\beta)\right\|^2 \left\|\Delta \hat{u}_{i_b} - \Delta u_{i_b}^*\right\|^2. \tag{7.51}$$

According to (7.19) and (7.40), and the mean-value theorem, the optimal incremental control is rewritten as

$$\Delta u_{i_b}^* = -\beta \tanh\left(\frac{1}{2\beta}B_i^\top \nabla \Phi_i^\top W_i^*\right) - \epsilon_{\Delta u_i^*}, \tag{7.52}$$

where $\epsilon_{\Delta u_i^*} = \frac{1}{2}(1 - \tanh^2(\eta_i))B_i^\top \nabla \epsilon_i$, and $\eta_i \in \mathbb{R}$ is chosen between $\frac{1}{2\beta}B_i^\top \nabla \Phi_i^\top W_i^*$ and $\frac{1}{2\beta}B_i^\top \nabla V_i^*$. According to $\|\nabla \epsilon_i\| \leq b_{\epsilon ei}$ in Assumption 7.1, $\left\|\epsilon_{\Delta u_i^*}\right\| \leq \frac{1}{2}\|B_i\| b_{\epsilon ei}$ holds. Then, by combining (7.45) with (7.52), we get

$$\Delta \hat{u}_{i_b} - \Delta u_{i_b}^* = \beta(\tanh(\mathscr{G}_i^*) - \tanh(\hat{\mathscr{G}}_i)) + \epsilon_{\Delta u_i^*}. \tag{7.53}$$

where $\mathscr{G}_i^* = \frac{1}{2\beta} B_i^\top \nabla \Phi_i^\top W_i^*$, and $\hat{\mathscr{G}}_i = \frac{1}{2\beta} B_i^\top \nabla \Phi_i^\top \hat{W}$. Based on (7.19) and (7.45), the Taylor series of $\tanh(\mathscr{G}_i^*)$ follows

$$\tanh(\mathscr{G}_i^*) = \tanh\left(\hat{\mathscr{G}}_i\right) + \frac{\partial \tanh\left(\hat{\mathscr{G}}_i\right)}{\partial \hat{\mathscr{G}}_i}(\mathscr{G}_i^* - \hat{\mathscr{G}}_i) + \mathcal{O}((\mathscr{G}_i^* - \hat{\mathscr{G}}_i)^2)$$

$$= \tanh\left(\hat{\mathscr{G}}_i\right) - \frac{1}{2\beta}(1 - \tanh^2(\hat{\mathscr{G}}_i))B_i^\top \nabla \Phi_i^\top \tilde{W}_i + \mathcal{O}((\mathscr{G}_i^* - \hat{\mathscr{G}}_i)^2),$$

$$(7.54)$$

where $\mathcal{O}((\mathscr{G}_i^* - \hat{\mathscr{G}}_i)^2)$ is a higher-order term of the Taylor series. By following [106, Lemma 1], this higher-order term is bounded as

$$\left\|\mathcal{O}((\mathscr{G}_i^* - \hat{\mathscr{G}}_i)^2)\right\| \le 2 + \frac{1}{\beta}\|B_i\| b_{\Phi_{ei}} \left\|\tilde{W}_i\right\|. \qquad (7.55)$$

Based on (7.54), we rewrite (7.53) as

$$\Delta \hat{u}_{i_b} - \Delta u_{i_b}^* = \beta(\tanh(\mathscr{G}_i^*) - \tanh\left(\hat{\mathscr{G}}_i\right)) + \epsilon_{\Delta u_i^*}$$

$$= -\frac{1}{2}(1 - \tanh^2(\hat{\mathscr{G}}_i))B_i^\top \nabla \Phi_i^\top \tilde{W}_i + \beta\mathcal{O}((\mathscr{G}_i^* - \hat{\mathscr{G}}_i)^2) + \epsilon_{\Delta u_i^*}.$$

$$(7.56)$$

Then, by combining (7.55) with (7.56), and given that $\left\|1 - \tanh^2(\hat{\mathscr{G}}_i)\right\| \le 2$, $\left\|\Delta \hat{u}_{i_b} - \Delta u_{i_b}^*\right\|^2$ in (7.51) follows

$$\left\|\Delta \hat{u}_{i_b} - \Delta u_{i_b}^*\right\|^2 \le 3\beta^2 \left\|\mathcal{O}((\mathscr{G}_i^* - \hat{\mathscr{G}}_i)^2)\right\|^2 + 3\left\|\epsilon_{\Delta u_i^*}\right\|^2$$

$$+ 3\left\|-\frac{1}{2}(1 - \tanh^2(\hat{\mathscr{G}}_i))B_i^\top \nabla \Phi_i^\top \tilde{W}_i\right\|^2 \qquad (7.57)$$

$$\le 6\|B_i\|^2 b_{\Phi_{ei}}^2 \left\|\tilde{W}_i\right\|^2 + 12\beta^2 + \frac{3}{4}\|B_i\|^2 b_{\epsilon_{ei}}^2$$

$$+ 12\beta \|B_i\| b_{\Phi_{ei}} \left\|\tilde{W}_i\right\|.$$

Based on (7.34), (7.40), Assumption 7.1, and the fact that $\|W_i^*\| \le b_{W_i^*}$, $\left\|\tanh^{-1}(\Delta u_{i_b}^*/\beta)\right\|^2$ in (7.51) follows

$$\left\|\tanh^{-1}\left(\Delta u_{i_b}^*/\beta\right)\right\|^2 = \left\|\frac{1}{4\beta^2}\nabla V_i^{*\top} B_i B_i^\top \nabla V_i^*\right\|$$

$$\le \frac{1}{4\beta^2}\|B_i\|^2 b_{\Phi_{ei}}^2 b_{W_i^*}^2 + \frac{1}{4\beta^2}b_{\epsilon_{ei}}^2 \|B_i\|^2 \qquad (7.58)$$

$$+ \frac{1}{2\beta^2}\|B_i\|^2 b_{\Phi_{ei}} b_{\epsilon_{ei}} b_{W_i^*}.$$

Using (7.57) and (7.58), (7.51) reads

$$- 2\beta \tanh^{-1}\left(\Delta u_{i_b}^*/\beta\right)\left(\Delta \hat{u}_{i_b} - \Delta u_{i_b}^*\right) \leq \frac{1}{4}\left\|B_i\right\|^2 b_{\Phi_{ei}}^2 b_{W_i^*}^2$$

$$+ \frac{1}{4}b_{\epsilon_{ei}}^2\left\|B_i\right\|^2 + \frac{1}{2}\left\|B_i\right\|^2 b_{\Phi_{ei}}b_{\epsilon_{ei}}b_{W_i^*} + 6\left\|B_i\right\|^2 b_{\Phi_{ei}}^2\left\|\tilde{W}_i\right\|^2 \qquad (7.59)$$

$$+ 12\beta^2 + \frac{3}{4}\left\|B_i\right\|^2 b_{\epsilon_{ei}}^2 + 12\beta\left\|B_i\right\| b_{\Phi_{ei}}\left\|\tilde{W}_i\right\|.$$

Substituting (7.59) into (7.50), finally the first term $\dot{L}_{i_1}$ follows

$$\dot{L}_{i_1} \leq - e_i^\top Q_i e_i - (\bar{\xi}_{oi}^2 - \xi_i^\top \xi_i) - \left[\beta\tanh^{-1}\left(\Delta u_{i_b}^*/\beta\right) + \xi_i\right]^2$$

$$+ \frac{3}{4}\left\|B_i\right\|^2 b_{\Phi_{ei}}^2 b_{W_i^*}^2 + \frac{3}{4}b_{\epsilon_{ei}}^2\left\|B_i\right\|^2 + \frac{3}{2}\left\|B_i\right\|^2 b_{\Phi_{ei}}b_{\epsilon_{ei}}b_{W_i^*} + 12\beta^2 + \frac{3}{4}\left\|B_i\right\|^2 b_{\epsilon_{ei}}^2$$

$$+ 6\left\|B_i\right\|^2 b_{\Phi_{ei}}^2\left\|\tilde{W}_i\right\|^2 + 12\beta\left\|B_i\right\| b_{\Phi_{ei}}\left\|\tilde{W}_i\right\|.$$

$$(7.60)$$

As for the second term $\dot{L}_W = \frac{1}{2}\tilde{W}_i^\top \Gamma_i^{-1}\tilde{W}_i$, based on (7.44) and Theorem 1 in our previous work [59], it follows

$$\dot{L}_{i_2} \leq - \tilde{W}_i^\top \mathcal{Y}_i \tilde{W}_i + \tilde{W}_i^\top \epsilon_{\tilde{W}_i}. \qquad (7.61)$$

where $\mathcal{Y}_i = \sum_{l=1}^{P_i} k_{e_i} Y_{i_l} Y_{i_l}^\top \in \mathbb{R}^{N_i \times N_i}$, and $\epsilon_{\tilde{W}_i} = -k_{t_i} Y_i \epsilon_{h_i} - \sum_{l=1}^{P_i} k_{e_i} Y_{i_l} \epsilon_{h_{il}} \in \mathbb{R}^{N_i}$. The boundness of Y_i and ϵ_{h_i} results in bounded $\epsilon_{\tilde{W}_i}$. Thus, there exists $\bar{\epsilon}_{\tilde{W}_i} \in \mathbb{R}^+$ such that $\left\|\epsilon_{\tilde{W}_i}\right\| \leq \bar{\epsilon}_{\tilde{W}_i}$. According to Assumption 7.2, $\mathcal{Y}_i$ is positive definite. Thus, (7.61) could be rewritten as

$$\dot{L}_{i_2} \leq - \lambda_{\min}(\mathcal{Y}_i)\left\|\tilde{W}_i\right\|^2 - \bar{\epsilon}_{\tilde{W}_i}\left\|\tilde{W}_i\right\|. \qquad (7.62)$$

Finally, as for $\dot{L}_i$, substituting (7.60) and (7.61) into (7.47), we get

$$\dot{L}_i \leq -\mathcal{A}_i - \mathcal{B}_i\left\|\tilde{W}_i\right\|^2 + \mathcal{C}_i\left\|\tilde{W}_i\right\| + \mathcal{D}_i, \qquad (7.63)$$

where $\mathcal{A}_i = e_i^\top Q_i e_i + (\bar{\xi}_{oi}^2 - \xi_i^\top \xi_i) + \left[\beta\tanh^{-1}\left(\Delta u_{i_b}^*/\beta\right) + \xi_i\right]^2$, $\mathcal{B}_i = \lambda_{\min}(\mathcal{Y}_i) - 6\left\|B_i\right\|^2 b_{\Phi_{ei}}^2$, $\mathcal{C}_i = 12\beta\left\|B_i\right\| b_{\Phi_{ei}} + \bar{\epsilon}_{\tilde{W}_i}$, and $\mathcal{D}_i = \frac{3}{4}\left\|B_i\right\|^2 b_{\Phi_{ei}}^2 b_{W_i^*}^2 + \frac{3}{2}b_{\epsilon_{ei}}^2\left\|B_i\right\|^2 + \frac{3}{2}\left\|B_i\right\|^2 b_{\Phi_{ei}}b_{\epsilon_{ei}}b_{W_i^*} + 12\beta^2$. Let the parameters be chosen such that $\mathcal{B}_i > 0$. Since $\mathcal{A}_i$ is positive definite, the Lyapunov derivative (7.63) is negative if

$$\left\|\tilde{W}_i\right\| > \frac{\mathcal{C}_i}{2\mathcal{B}_i} + \sqrt{\frac{\mathcal{C}_i^2}{4\mathcal{B}_i^2} + \frac{\mathcal{D}_i}{\mathcal{B}_i}}. \qquad (7.64)$$

Thus, the weight learning error of the critic agent converges to the residual set

$$\tilde{\Omega}_{\tilde{W}_i} = \left\{ \tilde{W}_i \middle| \left\| \tilde{W}_i \right\| \leq \frac{\mathcal{C}_i}{2\mathcal{B}_i} + \sqrt{\frac{\mathcal{C}_i^2}{4\mathcal{B}_i^2} + \frac{\mathcal{D}_i}{\mathcal{B}_i}} \right\}. \tag{7.65}$$

Bibliography

[1] Murad Abu-Khalaf, Jie Huang, and Frank L Lewis. *Nonlinear H2/H-Infinity Constrained Feedback Control: A Practical Design Approach Using Neural Networks.* Springer Science & Business Media, 2006.

[2] Murad Abu-Khalaf and Frank L Lewis. Nearly optimal control laws for nonlinear systems with saturating actuators using a neural network hjb approach. *Automatica*, 41(5):779–791, 2005.

[3] Paul Acquatella, E van Kampen, and Qi Ping Chu. Incremental backstepping for robust nonlinear flight control. *Proceedings of the EuroGNC*, 2013, 2013.

[4] Aaron D Ames, Kevin Galloway, Koushil Sreenath, and Jessy W Grizzle. Rapidly exponentially stabilizing control lyapunov functions and hybrid zero dynamics. *IEEE Transactions on Automatic Control*, 59(4):876–891, 2014.

[5] Aaron D Ames, Xiangru Xu, Jessy W Grizzle, and Paulo Tabuada. Control barrier function based quadratic programs for safety critical systems. *IEEE Transactions on Automatic Control*, 62(8):3861–3876, 2016.

[6] Brian Anderson. Exponential stability of linear equations arising in adaptive identification. *IEEE Transactions on Automatic Control*, 22(1):83–88, 1977.

[7] Michael Athans and Peter L Falb. *Optimal control: an introduction to the theory and its applications.* Courier Corporation, 2013.

[8] Nina Karlovna Bary. *A treatise on trigonometric series.* Elsevier, 2014.

[9] Thomas Beckers, Jonas Umlauft, Dana Kulic, and Sandra Hirche. Stable gaussian process based tracking control of lagrangian systems. In *2017 IEEE 56th Annual Conference on Decision and Control (CDC)*, pages 5180–5185, 2017.

[10] DIMITRI P BERTSEKAS. Approximate dynamic programming. 2012.

[11] Shubhendu Bhasin, Rushikesh Kamalapurkar, Huyen T Dinh, and Warren E Dixon. Robust identification-based state derivative estimation for nonlinear systems. *IEEE Transactions on Automatic Control*, 58(1):187–192, 2012.

[12] Shubhendu Bhasin, Rushikesh Kamalapurkar, Marcus Johnson, Kyriakos G Vamvoudakis, Frank L Lewis, and Warren E Dixon. A novel actor–critic–identifier architecture for approximate optimal control of uncertain nonlinear systems. *Automatica*, 49(1):82–92, 2013.

[13] Franco Blanchini and Stefano Miani. *Set-theoretic methods in control*. Springer, 2008.

[14] Joschka Boedecker, Jost Tobias Springenberg, Jan Wülfing, and Martin Riedmiller. Approximate real-time optimal control based on sparse gaussian process models. In *2014 IEEE Symposium on Adaptive Dynamic Programming and Reinforcement Learning (ADPRL)*, pages 1–8, 2014.

[15] Stephen Boyd and Sosale Shankara Sastry. Necessary and sufficient conditions for parameter convergence in adaptive control. *Automatica*, 22(6):629–639, 1986.

[16] Lucian Buşoniu, Tim de Bruin, Domagoj Tolić, Jens Kober, and Ivana Palunko. Reinforcement learning for control: Performance, stability, and deep approximators. *Annual Reviews in Control*, 46:8–28, 2018.

[17] Eduardo F Camacho and Carlos Bordons Alba. *Model predictive control*. Springer science & business media, 2013.

[18] Luis Rodolfo Garcia Carrillo and Kyriakos G Vamvoudakis. Deep-learning tracking for autonomous flying systems under adversarial inputs. *IEEE Transactions on Aerospace and Electronic Systems*, 56(2):1444–1459, 2019.

[19] Pyung Hun Chang and Je Hyung Jung. A systematic method for gain selection of robust pid control for nonlinear plants of second-order controller canonical form. *IEEE Transactions on Control Systems Technology*, 17(2):473–483, 2008.

[20] Wen-Hua Chen, Jun Yang, Lei Guo, and Shihua Li. Disturbance-observer-based control and related methods—an overview. *IEEE Transactions on Industrial Electronics*, 63(2):1083–1095, 2015.

[21] Girish V Chowdhary. *Concurrent Learning for Convergence in Adaptive Control Without Persistency of Excitation*. PhD thesis, Georgia Institute of Technology, 2010.

[22] Richard Courant and David Hilbert. *Methods of Mathematical Physics*. Vol I. Wiley (Interscience), New York, 1953.

[23] Martin Ester, Hans-Peter Kriegel, Jörg Sander, and Xiaowei Xu. A density-based algorithm for discovering clusters in large spatial databases with noise. In *Kdd*, volume 96, pages 226–231, 1996.

[24] William Fedus, Prajit Ramachandran, Rishabh Agarwal, Yoshua Bengio, Hugo Larochelle, Mark Rowland, and Will Dabney. Revisiting fundamentals of experience replay. *arXiv preprint arXiv:2007.06700*, 2020.

[25] Bruce A Finlayson. *The method of weighted residuals and variational principles*. SIAM, 2013.

[26] Harley Flanders. Differentiation under the integral sign. *The American Mathematical Monthly*, 80(6):615–627, 1973.

[27] Gene F Franklin, J David Powell, and Michael L Workman. *Digital control of dynamic systems*. Addison-wesley Reading, MA, 1998.

[28] Shuzhi Sam Ge, Chang C Hang, Tong H Lee, and Tao Zhang. *Stable adaptive neural network control*. Springer Science & Business Media, 2013.

[29] Mingfang He. Data-driven approximated optimal control for chemical processes with state and input constraints. *Complexity*, 2019, 2019.

[30] Ali Heydari and Sivasubramanya N Balakrishnan. Finite-horizon control-constrained nonlinear optimal control using single network

adaptive critics. *IEEE Transactions on Neural Networks and Learning Systems*, 24(1):145–157, 2012.

[31] Cherie Ho, Jay Patrikar, Rogerio Bonatti, and Sebastian Scherer. Adaptive safety margin estimation for safe real-time replanning under time-varying disturbance. *arXiv preprint arXiv:2110.03119*, 2021.

[32] Paul W Holland and Roy E Welsch. Robust regression using iteratively reweighted least-squares. *Communications in Statistics-theory and Methods*, 6(9):813–827, 1977.

[33] TC Steve Hsia. A new technique for robust control of servo systems. *IEEE Transactions on Industrial Electronics*, 36(1):1–7, 1989.

[34] Tien C Hsia and LS Gao. Robot manipulator control using decentralized linear time-invariant time-delayed joint controllers. In *IEEE International Conference on Robotics and Automation*, pages 2070–2075, 1990.

[35] Weihuang Huang, Guoliang Li, Kian-Lee Tan, and Jianhua Feng. Efficient safe-region construction for moving top-k spatial keyword queries. In *Proceedings of the 21st ACM international conference on Information and knowledge management*, pages 932–941, 2012.

[36] Mrdjan Jankovic. Robust control barrier functions for constrained stabilization of nonlinear systems. *Automatica*, 96:359–367, 2018.

[37] Maolin Jin, Jinoh Lee, Pyung Hun Chang, and Chintae Choi. Practical nonsingular terminal sliding-mode control of robot manipulators for high-accuracy tracking control. *IEEE Transactions on Industrial Electronics*, 56(9):3593–3601, 2009.

[38] Xu Jin. Adaptive fixed-time control for mimo nonlinear systems with asymmetric output constraints using universal barrier functions. *IEEE Transactions on Automatic Control*, 64(7):3046–3053, 2018.

[39] Rushikesh Kamalapurkar, Lindsey Andrews, Patrick Walters, and Warren E Dixon. Model-based reinforcement learning for infinite-horizon approximate optimal tracking. *IEEE Transactions on Neural Networks and Learning Systems*, 28(3):753–758, 2016.

[40] Rushikesh Kamalapurkar, Huyen Dinh, Shubhendu Bhasin, and Warren E Dixon. Approximate optimal trajectory tracking for continuous-time nonlinear systems. *Automatica*, 51:40–48, 2015.

[41] Rushikesh Kamalapurkar, Benjamin Reish, Girish Chowdhary, and Warren E Dixon. Concurrent learning for parameter estimation using dynamic state-derivative estimators. *IEEE Transactions on Automatic Control*, 62(7):3594–3601, 2017.

[42] Rushikesh Kamalapurkar, Joel A Rosenfeld, and Warren E Dixon. Efficient model-based reinforcement learning for approximate online optimal control. *Automatica*, 74:247–258, 2016.

[43] Rushikesh Kamalapurkar, Patrick Walters, and Warren E Dixon. Model-based reinforcement learning for approximate optimal regulation. In *Control of Complex Systems*, pages 247–273. Elsevier, 2016.

[44] Hassan K Khalil and Jessy W Grizzle. *Nonlinear systems*. Prentice hall Upper Saddle River, NJ, 2002.

[45] Bahare Kiumarsi, Kyriakos G Vamvoudakis, Hamidreza Modares, and Frank L Lewis. Optimal and autonomous control using reinforcement learning: A survey. *IEEE Transactions on Neural Networks and Learning Systems*, 29(6):2042–2062, 2017.

[46] William Koch, Renato Mancuso, Richard West, and Azer Bestavros. Reinforcement learning for uav attitude control. *ACM Transactions on Cyber-Physical Systems*, 3(2):1–21, 2019.

[47] Petar Kormushev, Sylvain Calinon, and Darwin G Caldwell. Reinforcement learning in robotics: Applications and real-world challenges. *Robotics*, 2(3):122–148, 2013.

[48] Shreyas Kousik, Sean Vaskov, Fan Bu, Matthew Johnson-Roberson, and Ram Vasudevan. Bridging the gap between safety and real-time performance in receding-horizon trajectory design for mobile robots. *The International Journal of Robotics Research*, 39(12):1419–1469, 2020.

[49] S Narendra Kumpati and Parthasarathy Kannan. Identification and control of dynamical systems using neural networks. *IEEE Transactions on Neural Networks*, 1(1):4–27, 1990.

[50] Sampo Kuutti, Richard Bowden, Yaochu Jin, Phil Barber, and Saber Fallah. A survey of deep learning applications to autonomous vehicle control. *IEEE Transactions on Intelligent Transportation Systems*, 2020.

[51] Steven M LaValle. *Planning algorithms*. Cambridge University Press, 2006.

[52] Steven J Leon, Ion Bica, and Tiina Hohn. *Linear algebra with applications*. Prentice Hall Upper Saddle River, NJ, 1998.

[53] Arie Levant. Robust exact differentiation via sliding mode technique. *Automatica*, 34(3):379–384, 1998.

[54] Frank L Lewis, Darren M Dawson, and Chaouki T Abdallah. *Robot manipulator control: theory and practice*. CRC Press, 2003.

[55] Frank L Lewis, Kai Liu, and Aydin Yesildirek. Neural net robot controller with guaranteed tracking performance. *IEEE Transactions on Neural Networks*, 6(3):703–715, 1995.

[56] Frank L Lewis and Draguna Vrabie. Reinforcement learning and adaptive dynamic programming for feedback control. *IEEE circuits and systems magazine*, 9(3):32–50, 2009.

[57] FW Lewis, Suresh Jagannathan, and Aydin Yesildirak. *Neural network control of robot manipulators and non-linear systems*. CRC Press, 2020.

[58] Cong Li, Fangzhou Liu, Yongchao Wang, and Martin Buss. Concurrent learning-based adaptive control of an uncertain robot manipulator with guaranteed safety and performance. *IEEE Transactions on Systems, Man, and Cybernetics: Systems*, 2021.

[59] Cong Li, Yongchao Wang, Fangzhou Liu, Qingchen Liu, and Martin Buss. Model-free incremental adaptive dynamic programming based approximate robust optimal regulation. *International Journal of Robust and Nonlinear Control*, 2022.

[60] Yongming Li, Yanjun Liu, and Shaocheng Tong. Observer-based neuro-adaptive optimized control of strict-feedback nonlinear systems with state constraints. *IEEE Transactions on Neural Networks and Learning Systems*, 2021.

[61] Yongming Li, Kangkang Sun, and Shaocheng Tong. Observer-based adaptive fuzzy fault-tolerant optimal control for siso nonlinear systems. *IEEE Transactions on Cybernetics*, 49(2):649–661, 2018.

[62] Feng Lin and Robert D Brandt. An optimal control approach to robust control of robot manipulators. *IEEE Transactions on Robotics and Automation*, 14(1):69–77, 1998.

[63] Feng Lin, Robert D Brandt, and Jing Sun. Robust control of nonlinear systems: Compensating for uncertainty. *International Journal of Control*, 56(6):1453–1459, 1992.

[64] Derong Liu, Shan Xue, Bo Zhao, Biao Luo, and Qinglai Wei. Adaptive dynamic programming for control: a survey and recent advances. *IEEE Transactions on Systems, Man, and Cybernetics: Systems*, 2020.

[65] Sikang Liu, Michael Watterson, Kartik Mohta, Ke Sun, Subhrajit Bhattacharya, Camillo J Taylor, and Vijay Kumar. Planning dynamically feasible trajectories for quadrotors using safe flight corridors in 3-d complex environments. *IEEE Robotics and Automation Letters*, 2(3):1688–1695, 2017.

[66] Yanbin Liu, Weichao Sun, and Huijun Gao. High precision robust control for periodic tasks of linear motor via b-spline wavelet neural network observer. *IEEE Transactions on Industrial Electronics*, 2021.

[67] Fangchao Luo, Bo Zhao, and Derong Liu. Event-triggered decentralized optimal fault tolerant control for mismatched interconnected nonlinear systems through adaptive dynamic programming. *Optimal Control Applications and Methods*, 2021.

[68] Teppo Luukkonen. Modelling and control of quadcopter. *Independent research project in applied mathematics, Espoo*, 22:22, 2011.

[69] Anirudha Majumdar and Russ Tedrake. Funnel libraries for real-time robust feedback motion planning. *The International Journal of Robotics Research*, 36(8):947–982, 2017.

[70] Ian M Mitchell. Comparing forward and backward reachability as tools for safety analysis. In *International Workshop on Hybrid Systems: Computation and Control*, pages 428–443, 2007.

[71] Ian M Mitchell, Alexandre M Bayen, and Claire J Tomlin. A time-dependent hamilton-jacobi formulation of reachable sets for continuous dynamic games. *IEEE Transactions on Automatic Control*, 50(7):947–957, 2005.

[72] Hamidreza Modares and Frank L Lewis. Optimal tracking control of nonlinear partially-unknown constrained-input systems using integral reinforcement learning. *Automatica*, 50(7):1780–1792, 2014.

[73] Jing Na, Guido Herrmann, and Kyriakos G Vamvoudakis. Adaptive optimal observer design via approximate dynamic programming. In *2017 American Control Conference (ACC)*, pages 3288–3293, 2017.

[74] Jing Na, Bin Wang, Guang Li, Siyuan Zhan, and Wei He. Nonlinear constrained optimal control of wave energy converters with adaptive dynamic programming. *IEEE Transactions on Industrial Electronics*, 66(10):7904–7915, 2018.

[75] Zhen Ni, Haibo He, and Jinyu Wen. Adaptive learning in tracking control based on the dual critic network design. *IEEE Transactions on Neural Networks and Learning Systems*, 24(6):913–928, 2013.

[76] Jorge Nocedal and Stephen Wright. *Numerical optimization*. Springer Science & Business Media, 2006.

[77] Warren B Powell. *Approximate Dynamic Programming: Solving the curses of dimensionality*. John Wiley & Sons, 2007.

[78] John O Rawlings, Sastry G Pantula, and David A Dickey. *Applied regression analysis: a research tool*. Springer Science & Business Media, 2001.

[79] Benjamin Recht. A tour of reinforcement learning: The view from continuous control. *Annual Review of Control, Robotics, and Autonomous Systems*, 2:253–279, 2019.

[80] Sayan Basu Roy and Shubhendu Bhasin. Robustness analysis of initial excitation based adaptive control. In *2019 IEEE 58th Conference on Decision and Control (CDC)*, pages 7055–7062, 2019.

[81] Hamid Sadeghian, Luigi Villani, Mehdi Keshmiri, and Bruno Siciliano. Task-space control of robot manipulators with null-space compliance. *IEEE Transactions on Robotics*, 30(2):493–506, 2013.

[82] Matteo Saveriano and Dongheui Lee. Learning barrier functions for constrained motion planning with dynamical systems. *arXiv preprint arXiv:2003.11500*, 2020.

[83] Tom Schaul, John Quan, Ioannis Antonoglou, and David Silver. Prioritized experience replay. *arXiv preprint arXiv:1511.05952*, 2015.

[84] Larry Schumaker. *Spline functions: basic theory*. Cambridge University Press, 2007.

[85] Yun Shen, Wilhelm Stannat, and Klaus Obermayer. Risk-sensitive markov control processes. *SIAM Journal on Control and Optimization*, 51(5):3652–3672, 2013.

[86] Yuri Shtessel, Christopher Edwards, Leonid Fridman, and Arie Levant. *Sliding mode control and observation*. Springer, 2014.

[87] P Simplicio, MD Pavel, E Van Kampen, and QP Chu. An acceleration measurements-based approach for helicopter nonlinear flight control using incremental nonlinear dynamic inversion. *Control Engineering Practice*, 21(8):1065–1077, 2013.

[88] Sumeet Singh, Benoit Landry, Anirudha Majumdar, Jean-Jacques Slotine, and Marco Pavone. Robust feedback motion planning via contraction theory. *The International Journal of Robotics Research*, 2019.

[89] Eduardo D Sontag. Smooth stabilization implies coprime factorization. *IEEE Transactions on Automatic Control*, 34(4):435–443, 1989.

[90] Eduardo D Sontag and Yuan Wang. On characterizations of the input-to-state stability property. *Systems & Control Letters*, 24(5):351–359, 1995.

[91] Wynn C Stirling. *Satisficing Games and Decision Making: with applications to engineering and computer science*. Cambridge University Press, 2003.

[92] Jingliang Sun and Chunsheng Liu. Disturbance observer-based robust missile autopilot design with full-state constraints via adaptive dynamic programming. *Journal of the Franklin Institute*, 355(5):2344–2368, 2018.

[93] Richard S Sutton and Andrew G Barto. *Reinforcement learning: An introduction*. MIT Press, 2018.

[94] Gang Tao. *Adaptive control design and analysis*. John Wiley & Sons, 2003.

[95] Keng Peng Tee and Shuzhi Sam Ge. Control of nonlinear systems with full state constraint using a barrier lyapunov function. In *Proceedings of the 48h IEEE Conference on Decision and Control (CDC) held jointly with 2009 28th Chinese Control Conference*, pages 8618–8623, 2009.

[96] Keng Peng Tee, Shuzhi Sam Ge, and Eng Hock Tay. Barrier lyapunov functions for the control of output-constrained nonlinear systems. *Automatica*, 45(4):918–927, 2009.

[97] Shaocheng Tong, Kangkang Sun, and Shuai Sui. Observer-based adaptive fuzzy decentralized optimal control design for strict-feedback nonlinear large-scale systems. *IEEE Transactions on Fuzzy Systems*, 26(2):569–584, 2017.

[98] Kyriakos G Vamvoudakis and Frank L Lewis. Online actor–critic algorithm to solve the continuous-time infinite horizon optimal control problem. *Automatica*, 46(5):878–888, 2010.

[99] Kyriakos G Vamvoudakis, Draguna Vrabie, and Frank L Lewis. Online adaptive algorithm for optimal control with integral reinforcement learning. *International Journal of Robust and Nonlinear Control*, 24(17):2686–2710, 2014.

[100] John Villadsen and Michael L Michelsen. *Solution of differential equation models by polynomial approximation*. Prentice-Hall Englewood Cliffs, NJ, 1978.

[101] Li Wang, Evangelos A Theodorou, and Magnus Egerstedt. Safe learning of quadrotor dynamics using barrier certificates. In *2018 IEEE International Conference on Robotics and Automation (ICRA)*, pages 2460–2465, 2018.

[102] Greg Welch and Gary Bishop. An introduction to the kalman filter, 1995.

[103] Wei Xiao and Calin Belta. High order control barrier functions. *IEEE Transactions on Automatic Control*, 2021.

[104] Wei Xiao, G Christos Cassandras, and Calin Belta. Safety-critical optimal control for autonomous systems. *Journal of Systems Science and Complexity*, 34(5):1723–1742, 2021.

[105] Xiangru Xu. Constrained control of input–output linearizable systems using control sharing barrier functions. *Automatica*, 87:195–201, 2018.

[106] Xiong Yang, Derong Liu, Hongwen Ma, and Yancai Xu. Online approximate solution of hji equation for unknown constrained-input nonlinear continuous-time systems. *Information Sciences*, 328:435–454, 2016.

[107] Yongliang Yang, Da-Wei Ding, Haoyi Xiong, Yixin Yin, and Donald C Wunsch. Online barrier-actor-critic learning for h-∞ control with full-state constraints and input saturation. *Journal of the Franklin Institute*, 2019.

[108] Kamal Youcef-Toumi and S-T Wu. Input/output linearization using time delay control. *Journal of dynamic systems, measurement, and control*, 114(1):10–19, 1992.

[109] Huaguang Zhang, Lili Cui, Xin Zhang, and Yanhong Luo. Data-driven robust approximate optimal tracking control for unknown general nonlinear systems using adaptive dynamic programming method. *IEEE Transactions on Neural Networks*, 22(12):2226–2236, 2011.

[110] Kun Zhang, Rong Su, Huaguang Zhang, and Yunlin Tian. Adaptive resilient event-triggered control design of autonomous vehicles with an iterative single critic learning framework. *IEEE Transactions on Neural Networks and Learning Systems*, 2021.

[111] Bo Zhao and Derong Liu. Event-triggered decentralized tracking control of modular reconfigurable robots through adaptive dynamic programming. *IEEE Transactions on Industrial Electronics*, 67(4):3054–3064, 2019.

[112] Pan Zhao, Arun Lakshmanan, Kasey Ackerman, Aditya Gahlawat, Marco Pavone, and Naira Hovakimyan. Tube-certified trajectory tracking for nonlinear systems with robust control contraction metrics. *IEEE Robotics and Automation Letters*, 7(2):5528–5535, 2022.

[113] Wanbing Zhao, Hao Liu, and Frank L Lewis. Robust formation control for cooperative underactuated quadrotors via reinforcement learning. *IEEE Transactions on Neural Networks and Learning Systems*, 2020.

[114] Ye Zhou, Erik-Jan van Kampen, and QiPing Chu. Nonlinear adaptive flight control using incremental approximate dynamic programming and output feedback. *Journal of Guidance, Control, and Dynamics*, 40(2):493–496, 2016.

[115] Ye Zhou, Erik-Jan van Kampen, and Qi Ping Chu. Incremental model based online dual heuristic programming for nonlinear adaptive control. *Control Engineering Practice*, 73:13–25, 2018.

[116] Ye Zhou, Erik-Jan Van Kampen, and Qiping Chu. Incremental model based online heuristic dynamic programming for nonlinear adaptive tracking control with partial observability. *Aerospace Science and Technology*, 105:106013, 2020.

[117] Antoni Zygmund. *Trigonometric series*. Cambridge University Press, 2002.

Index

For Product Safety Concerns and Information please contact our EU
representative GPSR@taylorandfrancis.com
Taylor & Francis Verlag GmbH, Kaufingerstraße 24, 80331 München, Germany